"Why do engineers prefer GLP FE/EIT study guides?"

Reader remarks from the GLP in-basket....

"I greatly appreciate your coming out with such an excellent *FE/EIT Discipline Review*. I walked into the exam feeling prepared and confident...and most important, I passed!"
—*Bill Johnston, Raleigh, NC*

"After looking at several other books, I am convinced your *Fundamentals of Engineering* is the best review around. It is very well organized, thorough, and concise. An outstanding book."
—*Paul Griesmer, Cleveland, OH*

"This is a time- and cost-effective tool. I easily passed the FE on my first attempt."
—*Omar Mureebe, Uniroyal Chemical Company, Middlebury, CT*

"In your FE book you cut to the chase and don't embellish it... Yours is the only review that doesn't scare the pants off people. It's exactly what I want for teaching my review course."
—*John Thorington*

"I was talking to a student a few hours ago and he said the morning problems on the FE/EIT were like the sample problems in your book—verbatim. It's the best review I've ever seen. When we started using your review in 1994, our students' pass rates went up 20%! Now we will continue to use your book because we take professional licensing seriously."
—*Dr. Gary Rogers, Virginia Military Institute*

"Your FE study guide is very thorough, well-organized and easy to use. It was instrumental in helping me brush up on engineering material I had not seen for 5 years. I passed the FE the first time!"
—*Brian Campbell, Project Engr, Trailmaster Corp, Ft. Worth, TX*

"Thank you for the book! I think I passed the test because of it. Stuff I hadn't studied for more than 15 years came back to me clearly. The two practice exams are what helped me the most."
—*Juan Alfaro, College Park, MD*

"Three years out of school and all I needed to pass the FE exam, on the first try, was your review book! I have recommended it to everyone I know. I look forward to using your PE reference."
—*David Robins, Syracuse, NY*

"Your book is great! Very easy to use. Although I only had time to skim through it, it helped me review the very basics I needed to know (i.e., Chemistry, Econ). I passed without a problem! Thanks! I'm keeping it as a handy reference."
—*Julie Volby, Civil Engineering Student, Univ of Minnesota, Minneapolis, MN*

"Your review material is perfect for the non-traditional 'night student.' It refreshes and restores many years of part-time study, without being overly exhaustive."
—*Jerome Bobak, Designer Mechanical Engineer, Grand Island, NY*

"This book provided all the necessary information for passing the FE exam. I was able to review for and *pass* the exam in only 3 weeks!"
—*Joseph Rozza, Recent Graduate, Orlando, FL*

Important Information

FE examination date: _____ time: _____

location: _____

_____

Examination Board Address: _____

_____

_____

phone: _____

Application was requested on this date: _____

Application was received on this date: _____

Application was accepted on this date: _____

This book belongs to: _____

_____

_____

_____

phone: _____

•FE/EIT•
Civil Engineering Review

5th edition

Merle C. Potter, Phd, PE—Editor

An Efficient Review for the Afternoon Discipline Test in Civil Engineering

From the professors who know it best...

Thomas F. Wolff	*Geotechnical*
Frank J. Hatfield	*Structural Analysis*
Thomas L. Maleck	*Transportation*
Francis McKelvey	*Airport Design*
Gilbert Baladi	*Pavement Design*
Susan J. Masten	*Environmental Engineering*
Merle C. Potter	*Hydrology*

published by:

GREAT LAKES PRESS

Okemos, Michigan   Wildwood, Missouri
P.O. Box 550, Wildwood MO 63040
Customer Service (636) 273-6016   www.glpbooks.com

International Standard Book Number 1-881018-07-5

Copyright © 2001 by Great Lakes Press, Inc.

All rights are reserved.
No part of this publication may be reproduced in any form or
by any means without prior written permission of the publisher.

All comments and inquiries should be addressed to:
    Great Lakes Press
    PO Box 550
    Wildwood, MO 63040-0550
    Phone (636) 273-6016
    Fax (636) 273-6086
    www.glpbooks.com
    custserv@glpbooks.com

Library of Congress Control Number: 2001091904

Printed in the USA by Sheridan Books, Inc. of Ann Arbor, Michigan.

10 9 8 7 6 5 4 3 2 1

Table of Contents

A Brief Outline of The FE/EIT Exam 3
State Boards of Registration Information 10

Passing the FE/EIT Exam

1. Geotechnical .... *Wolff* 17
   1.1 Weight–Mass–Volume Problems 17
   1.2 Relative Density 27
   1.3 Grain-Size Characteristics of Soils 28
   1.4 Atterberg Limits and Plasticity 31
   1.5 Stresses in Soil 32
   1.6 Water Flow in Soil 35
   1.7 Settlement of Saturated Clay 40
   1.8 Shear Strength of Soils 45
   1.9 Bearing Capacity of Soils 48
   1.10 Earth Pressure and Retaining Walls 50
   1.11 Slope Stability 52

2. Structural Analysis and Design .... *Hatfield* 55
   2.1 Analysis of Determinate Trusses 55
   2.2 Analysis of Determinate Beams 57
   2.3 Design in Reinforced Concrete 59
   2.4 Design in Structural Steel 61

3. Transportation .... *Maleck* 63
   3.1 Braking Distance 63
   3.3 Sight Distance for a Sag Vertical Curve 67
   3.4 Vertical Curve Elevations 69
   3.5 Horizontal Curves 71
   3.6 Superelevations 73
   3.7 Spirals 74

4. Airport Design .... *McKelvey* 75

5. Pavement Design .... *Baladi* 85

6. Environmental Engineering .... *Masten* 91
   6.1 Hardness Removal 91
   6.2 Grit Chamber Design 94
   6.4 Activated Sludge 96
   6.5 Fixed-film Processes 99
   6.6 Sludge Production 100
   6.7 Chemical Dosing 101
   6.8 Biochemical Oxygen Demand (BOD) 102
   6.9 Sewage Flow Ratio Curves 103
   6.10 Hydraulic-Elements Graph 104

7. Hydrology and Fluid Flow .... *Potter* 105

The FE/EIT Review for Civil Engineering

Civil Engineering Practice Exam 109

CE Practice Exam

Appendix A—NCEES Equation Summaries 141
Appendix B—Units and Conversions 153

Appendixes: Equations & Units

Preface

by Merle C. Potter

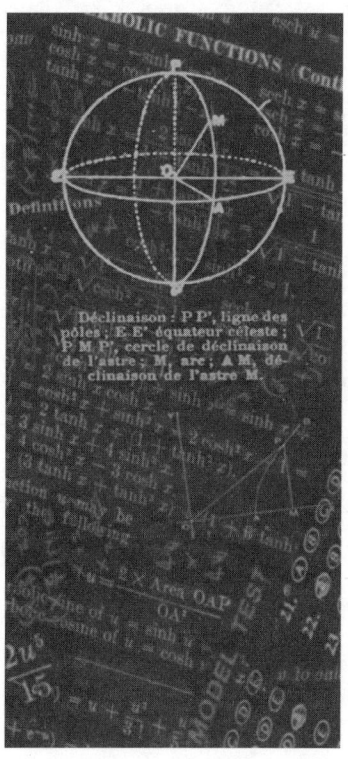

The FE/EIT Exam Format

The FE/EIT exam is a basic competency exam that consists of a General engineering test during the morning session, covering twelve subject areas required in most engineering curricula, and an afternoon session that is discipline-specific. The afternoon session will consist of one of six discipline-specific tests: Chemical, Civil, Electrical, Industrial, Mechanical, or General. The General test is intended for those engineers who do not fall into any of the five specific disciplines. Each examinee will take the morning General test and one of the afternoon tests.

Since there are over 30 engineering disciplines, examinees in over 25 of those disciplines will be expected to take the General test during the afternoon session. Most of the state boards will recognize a "pass" on any of the afternoon tests by any examinee. Consequently, it may be advisable for you to elect the afternoon General test even though you are a Civil, Electrical, Mechanical, Chemical, or Industrial engineer. Check with your state board if you are uncertain of your state's position.

Review Overview

This book was written as a review for the afternoon test of the FE/EIT exam in Civil engineering. Our strategy is to offer a short review of each subject and illustrate each subject with at least one example problem which is as exam-like as possible; this has been done for the subjects and primary equations included in the official NCEES Reference Handbook. Since the Handbook is the only book allowed into the examination, this review should cover nearly all problems tested on the afternoon session of the FE/EIT exam.

We offer a full 60-question practice test, with full solutions. This exam should help you decide if you should take the discipline test in your area or select the General test option.

NCEES Handbook

Many of the afternoon questions can be answered by referring to equations, tables, or charts included in the first 83 pages of the NCEES Reference Handbook —material that is primarily intended for the General tests. We recommend that you use our *Fundamentals of Engineering Review*, a book written to help you prepare for the morning General test, and much of the afternoon discipline test, as well as the afternoon General test. Call 1-800-837-0201 to order a copy. We also offer HP 48GX preprogrammed calculators and a companion 240-page *Jump Start the HP 48G/X* helpbook—highly effective aids to help you pass the FE/EIT exam.

Free CD Tests!

Finally, to further help you decide if you are ready to take the test in your discipline, we have included a coupon at the back cover of this book for a free CD which presents CE, ME, EE and IE practice exams in a user-friendly, rather attractive interactive environment. The CD will interpret the results of your test, suggesting if you are ready for the actual exam and indicating the score you obtained in each of the subject areas of the exam. It also gives you study strategy suggestions.

To obtain your copy of the free CD exams, simply fill out the coupon at the back of the book and drop it in the mail.

Best wishes in your study and on your test day!

Dr. Merle C. Potter, PE
Okemos, Michigan

Passing the FE/EIT Exam

This book presents an efficient review of the major Civil Engineering subjects tested in the afternoon session of the FE/EIT exam. GLP offers a companion book, *Fundamentals of Engineering Review*, which presents a succinct yet complete review of the 12 major topics of the morning General session, which you'll also have to take. You may order a copy of our General review by calling 1-800-837-0201.

A Brief Outline of This Review

This book contains:
- Short, succinct reviews of the important aspects of most of the Civil Engineering subject areas—just enough coverage to ensure that all-important pass! Some subject areas are presented with sample problems only.
- Sample problems that present the situational use of the more important equations presented in the NCEES Handbook.
- Complete solutions to each sample problem showing the details needed to arrive at a solution.
- A full 60-question practice test, with solutions.
- Mail-in coupon for *free* CD containing exams for 4 different disciplines, plus our unique study strategy software called Study-Director™.

In 1983, we at Great Lakes Press developed an FE/EIT review with the cooperation of a team of colleagues in response to our experience in coordinating FE/EIT review courses. The kind of well-planned material that an engineer naturally desires to use did not exist at the time. The options were either to pick and

Why We Created This FE Review

choose from a large, encyclopedia-like book that covers almost every topic in engineering, or to use material that didn't even attempt to cover the topics tested on the exam. It was either feast or famine, and it was all awkward—and often rather expensive! So we recruited popular lecturers from the university campus and prepared our first *Fundamentals of Engineering Review*.

With the announcement of the new exam format in 1996, we quickly pulled together a team of 20 professors to prepare discipline-specific reviews in time for the very first test in the new format.

On a daily basis we act as advocate for the test-taker in our involvement with engineering departments, associations, review courses and registration governing bodies. We are dedicated to making the entire licensing and registration process as reasonable as possible. We've responded faster than any other reviewing option to all of the many exam changes that keep coming at test-takers.

Our strategy for beating the Discipline-Specific format is a prime example of our leadership. We've prepared this review focusing solely on the equations, figures, and tables presented in the NCEES Reference Handbook. This seems to be the most reasonable way to prepare for the very many topics covered by discipline-specific tests. It is vital for effective familiarization not to get too detailed in one's study for such broad exams. Also, the Handbook is the only aid allowed into the exam besides a calculator. There will be questions that may not require the use of the equations from the Handbook; these will be based on general information from the respective courses. Such general information may not be included in this review.

Our "Pass or Your Money Back" Guarantee!

For your added confidence, with GLP you're guaranteed to pass! If you use our preps and fail the FE exam, send us your name & address, receipt, book & CD, and copy of failure notice (within 30 days of notice)—and we will issue you a full refund.

How to Become a Professional Engineer

To become registered as an engineer, a state may require that you:
1. Graduate from an ABET-accredited engineering program
2. Pass the *Fundamentals of Engineering* exam
3. Pass the *Principles and Practice of Engineering* (PE) exam after several years of engineering experience

Requirements vary from state to state, so you should obtain local guidelines and follow them carefully.

Why Get an Engineering License?

Registration is necessary if an engineer works as a consultant, and is highly recommended in certain industries—especially when one is, or hopes to be, in a management position.

How the FE Exam Is Scored

In the morning session, each of the 120 problems is worth <u>one-half</u> point. Thus a maximum score of 60 is possible. The afternoon session is different from the morning in that each problem is worth <u>one</u> point. With half the number of questions, the maximum possible score in the afternoon session is also 60.

The two-session total is 120 points, with both sessions having equal weight. A predetermined passing percentage is not established, nor is the exam graded on a curve. From recent history, a score of 60 (50% correct) would probably be a passing score. Each year the exact score is determined by statistical methods, so the score of 60 is an approximation.

Examination Format

The FE is composed of a four-hour general-subject morning session, followed by a one-hour break, then a four-hour discipline-specific afternoon session of 60 questions. The two sessions have a total of 180 4-part multiple-choice questions covering the subject areas listed below.

Morning Session

Subject Area	*Approximate Number*
Chemistry	11
Computers	6
Dynamics	10
Economics	5
Electrical Circuits	12
Ethics	5
Fluids	8
Materials Science	8
Mathematics	24
Mechanics of Materials (Solids)	8
Statics	12
Thermodynamics	11

Total Questions: 120

Subject Lists of the Discipline-Specific Afternoon Tests

General • Chemistry, Computers, Dynamics, Economics, Electrical, Ethics, Fluids, Material Science, Mathematics, Mechanics of Materials, Statics, Thermo

Civil • Construction, Enviro, Hydraulics, Hydrology, Numerical Methods, Legal, Soils, Foundations, Structural, Design, Surveying, Water Treatment, Waste Water

Mechanical • Controls, Computers, Systems, Energy Conversion, Power Plants, Turbomachines, Fluids, Heat Transfer, Materials, Instrumentation, Design, HVAC, Solids, Thermodynamics

Electrical • Analog, Communication, Numerical Methods, Computer Hardware, Software, Controls, Systems, Digital, Electromagnetics, Instrumentation, Networks, Power, Signal Processing, Solid State Devices

Chemical • Reactions, Thermodynamics, Numerical Methods, Heat Transfer, Mass Transfer, Energy, Pollution, Controls, Design, Economics, Equipment, Safety, and Transport Phenomena

Industrial • Design, Econ, Statistics, Facilities, Cost Analysis, Computer Modeling, Ergonomics, Management, Systems, Manufacturing, Material Handling, Optimization, Production, Productivity, Queuing, Simulation, Quality Control, Quality Management, Performance

How to Use the NCEES Handbook

A copy of the NCEES Handbook may be mailed to you sometime after you register. It is the only reference allowed in the exam room, so you'll want to get to know it. If you're not sent a copy by your state board, you can order one from GLP (by way of our website or call 1-800-837-0201). Your first impression of this handbook may leave you overwhelmed! It has a tremendous amount of equations, figures, and information that may, at first glance, seem unfamiliar to you. Indeed, afternoon Discipline test info is mixed in with morning General info. In our review we have used the NCEES Handbook to compile summaries of the more important equations from the most significant General subject areas, fitting in with the Morning Session and our recommended Afternoon strategy—this is Appendix A at the back of this book. We have attempted to adapt our nomenclature throughout to that of the NCEES Handbook.

We strongly urge you to have the official NCEES Reference Handbook by your side the entire time you study, and use a highlighter to identify the equations you most often use. Initially, you can use our Equation Summary Sheets and your own best judgment to highlight the key equations in the general part of the Handbook.

Familiarize yourself with the location of these equations so you can quickly locate them during the actual exam in the new handbook that will be given to you at the test site. If you do not prepare this way, you might become lost in the handbook during the exam.

Strategies for Study

The following strategies will help you prepare for the FE General Sessions:

1. Focus your review on the eight subjects that have the most exam questions: Chemistry, Dynamics, Electrical, Fluids, Math, Solids, Statics, and Thermo.
2. *Quickly* review the remaining four major subject areas: Computers, Economics, Ethics and Materials Science.
3. Spend the majority of your time reviewing material with which you are *most* familiar.
4. Study throughout your review with the NCEES Handbook in a targeted, realistic manner.

You may receive a copy of the NCEES Handbook when you register for the exam (some states provide them), or you can order a copy from GLP by calling 1-800-837-0201. You will not be allowed to bring this handbook (or any other printed material) with you into the exam. However, a clean copy of this same handbook will be given to you upon entering the exam site. The new handbook is particularly confusing. It includes about 80 pages that cover the General subjects (a.m. and p.m.) and 45 pages that cover the Discipline-Specific subjects. The trouble is that the a.m. and p.m. subjects are intermixed! ME's, for instance, find Heat Transfer covered early in the Handbook, hidden in with most of the General topics. If you elect to take the General Test, boldly mark the pages that cover the general subjects. In our book, we have provided most of the same equations that you will find in the NCEES Handbook for the major a.m. and General Test p.m. subjects. Then search through the Handbook and highlight those important equations. This will allow you to familiarize yourself with the location of the equations you need most, and you will be able to quickly locate those equations in the Handbook during the exam. You will find this particularly helpful since the Handbook is filled with plenty of extraneous material. If you do not train yourself in this way, you may well find it difficult to quickly locate the appropriate equations during the exam. Being familiar with the Handbook is perhaps the most important detail you should attend to.

Outline of a Good Study Program for Quick Progress

For a senior in an engineering college who has a busy schedule, we suggest 8 weeks as an ideal study period. During those 8 weeks, you must be willing to perform a fairly concentrated study. You should set aside blocks of 3 hours at least 2 days a week. In any case, we recommend that you study no less than 4 weeks for the FE exam. For those of you who have been away from this material for a while, base the length of your study period on the number of years you have been away from school and your own memory capability.

Perform an initial review during the first half of your selected study period (for engineering seniors, 4 weeks). But if you are trying to get through your review quickly, only the Practice Problems that have been starred (*) should be worked and studied in our *Fundamentals of Engineering Review* (General Test).

Two days prior to exam day, review all subject areas briefly, using your highlighted handbook of equations and tables. Be sure you can quickly find the equations you will use most often in the *unmarked* booklet you will be given on exam day.

The day before the exam, relax and go to bed early. Do *not* cram or perform any panic studying. By now, you are as prepared as you can be for the exam.

The morning of your test, get up early and have a light, healthy breakfast (and maybe some coffee!). Arrive at the exam site at least 20 minutes early. You need to allow time for parking and getting settled—be sure to bring some change to pay for parking! During the one-hour lunch break, it is best to plan to meet with a friend who can help you relax and get refreshed for the Afternoon Session. If there are no restaurants nearby, bring a bag lunch and eat outside on the lawn somewhere. Get some fresh air. Do *not* spend the entire hour reviewing. Try not to talk to other test-takers about how it is going for them. This can easily induce either insecurity or false confidence. If you understand engineering principles and have prepared well, after the dust and sweat of the test day clears you'll find you've passed!

Pacing Yourself During the FE Exam

The problems in the morning session are, for the most part, unrelated. Consequently, *two minutes* (on the average) can be spent on each problem. This makes fast recall *essential*, as time does not allow you to contemplate various methods of solution. But each problem is only worth one-half point—so do not fall into the trap of spending too much time on any one problem from the morning session.

In the afternoon session, you can spend an average of 4 minutes per individual question. Many questions in the afternoon session are related to a common problem, divided into 2 to 6 sub-problems. Thus you may maintain a good pace and still spend up to 20-25 minutes solving a problem set that is more difficult.

Process of Elimination

Sometimes the best way to find the right answer is to look for the wrong ones and cross them out. On questions which are difficult for you, wrong answers are often much easier to find than right ones!

Answers are seldom given with more than three significant figures, and may be given with two significant figures. The choice *closest* to your own solution should be selected.

There is no penalty for *guessing*. Use the *process of elimination* when guessing at an answer. If only one answer is negative and three answers are positive, eliminate the one odd answer from your guess. Also, when in doubt, work backwards and eliminate those answers that you believe are untrue, and then guess. By using a combination of methods, you greatly improve your odds of answering correctly.

Should I Guess?

Leave the last ten minutes of each session for making educated guesses. **Do not leave any answers blank** on your answer key. A guess *cannot* hurt you, it can only help you. Your score is based on the number of questions you answer correctly. An incorrect answer does not harm your score.

Place a question mark beside choices you are uncertain of, but seem correct. If time prevents you from reworking that problem, you will have at least identified your best guess.

Difficult Problems

If at first glance you know that a certain problem will require much time and is exceptionally difficult for you, make your best guess, then *skip right over it*. Be sure to mark the problem in a unique manner (we suggest that you circle the problem number) so that if time permits, you may come back to it. (Note: it is not possible to return to the Morning Session problems in the Afternoon Session.)

When you are working through a problem and decide to move on due to difficulty, be sure to write down in your test booklet your notes and conclusions up to that point in case you have time to return to it. Then make your best guess and circle the problem number to return to if you have time.

If you feel you know how to work a difficult problem and could answer it with more time, identify it by circling the entire problem, not just the problem number—this identifies it as a "most likely" candidate for your set-aside ten minutes of "guess" time.

Time-Saving Tips

Once you determine your answer, *always write the letter corresponding to the correct answer in the margin of the test booklet* beside the question. At the end of the page, you can then transfer all the answers from that page to the answer key at once. This will save you considerable time and help you maintain concentration as well!

Cross out choices that you have eliminated in your test booklet on the problems you will return to. Otherwise, you will have to reread them as you make your last-ditch deliberations.

Write Out Your Work

Feel free to write all over the writing space provided for you in your test booklet. Do not hesitate to work out a problem, no matter how simple it may be. Doing as much work as you can on paper will ease your mind and leave it less "cluttered"—and it will help you if you need to return to a problem. You may think you're saving time, but your exam performance is *not* improved when you work problems in your head.

Bring a Calculator —Ideally, an HP 48GX

You must take a silent calculator (it may be preprogrammed) into the exam. A calculator is essential when solving many problems. In fact, with the exception of a couple of states, the premier engineering calculator, the HP 48GX, is allowed into the exam (check with your state board). This calculator is a hand computer which has hundreds of basic equations and constants preprogrammed. If you need extra help with exam calculations, the HP 48GX will be very useful. We at GLP offer this calculator for sale at a substantial discount. The GX, with its large memory and removable card-slot, is of particular importance because cards are made for it especially for the FE and PE exams —we have these cards available. We also offer a manual to guide you through the steps to using this calculator effectively for the FE Exam. (The manufacturer's manual is difficult to use for even basic operations.) Our engineering-oriented manual, called *Jump Start the HP 48G/GX*, will have you performing typical calculations in a very short time. Call us at 1-800-837-0201 for ordering or information, or browse our website with its easy-to-use order form at www.glpbooks.com.

English vs. SI Units

Some questions can be worked using either English units or SI units in the morning session only. We recommend that all test takers prepare using only SI units, although some of our review material uses English units since that is what is used in the course. The answers from the array of choices will be the same using either set of units. The problems in this book are mostly in SI units. The afternoon Civil Engineering Test will use both sets of units in some of the subjects areas (e.g., Soils). A table of conversion factors is presented in Appendix B of this book.

Recommended Materials for FE/EIT Preparation

1. *FE/EIT Discipline Reviews for Civil, Mechanical or Electrical*, (this book); GLP
2. *Fundamentals of Engineering Review—General*; a complete review; GLP
3. *FE/EIT Quick Prep*—for General sessions; a quick review; GLP
3. HP 48GX Programmable Calculator
4. *Jump Start the HP 48G/GX*, a how-to manual from GLP
5. NCEES FE/EIT Sample Exams, for each Discipline
6. NCEES Reference Handbook

All these resources, and more, are available from GLP by calling 1-800-837-0201.

Again... "Pass or Your Money Back"!

Did we forget to mention that with GLP you're guaranteed to pass? It's true! If you use our preps and fail the FE or PE exam, send us your name & address, receipt, book & CD, and copy of failure notice (within 30 days of notice)—and we will issue you a full refund.

State Boards of Registration Information

All State Boards of Registration administer the National Council of Engineering Examiners and Surveyors (NCEES) uniform examination. The dates of the exams cover a span of three days in mid-April and three days in late October. The specific dates are selected by each State Board. To be accepted to take the FE exam, an applicant must apply well in advance. For information regarding the specific requirements in your state, contact your State Board's office. If contact information has changed from what we have listed here, your correct State Board information can be obtained from the Executive Director of NCEES, PO Box 1686, Clemson, SC 29633-1686, ph 803-654-6824, or www.NCEES.org. Any comments relating to the exam or the Reference Handbook should be addressed to NCEES .

The answers to the following questions are answered alongside each listing:

1. *Do you provide an NCEES Handbook for all registrants?*
2. *Do you allow CE's, ME's, EE's, IE's and ChemE's to take the afternoon General Exam?*
3. *Are advanced calculators, such as the HP48GX, allowed?*

NCEES Handbook at signup?	Can all take Gen'l Exam?	Advanced calculators allowed?	
Y	Y	Y	ALABAMA: State Board of Licensure for Professional Engineers, P. O. Box 304451, Montgomery 36130-4451. Executive Secretary, Telephone: (334) 242-5568, engineer@dsmd.dsmd.state.al.us.
Y	Y	Y	ALASKA: State Board of Registration for Engineers, Pouch D, Juneau 99811. Licensing Examiner, Telephone: (907) 465-2540, marcia_pappas@dced.state.ak.us, www.commerce.state.ak.us/occ/pael/htm.
N	N	Y	ARIZONA: State Board of Technical Registration, 1990 W. Camelback Rd., Suite 406, Phoenix 85015. Executive Director, Telephone: (602) 255-4053, btrlvd@yahoo.com, www.btr.state.az.us.
Y	Y	'Y'	ARKANSAS: State Board of Registration for Professional Engineers and Land Surveyors, P. O. Box 3750, Little Rock 72203. Secretary-Treasurer, Telephone: (501) 682-2824, joe.clements@mail.state.ar.us. www.state.ar.us/pels. (No pre-programmed calculator cards allowed.)
Y	Y	Y	CALIFORNIA: Board for Professional Engineers and Land Surveyors, 2535 Capitol Oaks Dr #300, Sacramento 95833. Executive Secretary, Telephone: (916) 263-2222, www.dca.ca.gov/pels.
'Y'	Y	Y	COLORADO: State Board of Registration for Professional Engineers, 1560 Broadway, Suite 1370, Denver 80202. Program Administrator, Telephone: (303) 894-7788, www.dora.state.co.us/engineers. (Handbooks avail. while they last.)
N	Y	Y	CONNECTICUT: State Board of Registration for Professional Engineers, The State Office Building, Rm 110, 165 Capitol Ave, Hartford 06106. Administrator, Telephone: (860) 713-6145.

	NCEES Book?	Gen'l Exam?	Calculators?

DELAWARE: Association of Professional Engineers, 56 W. Main St, Suite 208, Christina 19702. Executive Secretary, Telephone: (302) 368-6708, peggy@dape.org, www.dape.org. — Y Y Y

DISTRICT OF COLUMBIA: Board of Registration for Professional Engineers, 941 N. Capitol St., OPLA Rm 2200 Washington 20002. Executive Secretary, Telephone: (202) 442-4320. — Y Y Y

FLORIDA: Board of Professional Engineers, 1208 Hays St., Tallahassee 32301-0755. Executive Director, Telephone: (850) 521-0500, board@fbpe.org, www.fbpe.org. — N Y Y

GEORGIA: State Board of Registration for Professional Engineers, 237 Coliseum Dr., Macon, 31217-3858. Executive Director, Telephone: (912) 207-1450, pels@sos.state.ga.us, www.sos.state.ga.us/ebd–pels. — Y Y Y

GUAM: Territorial Board of Registration for Professional Engineers, Architects and Land Surveyors, Department of Public Works, Government of Guam, P. O. Box 2950, Agana 96911. Chairman, Telephone: (671) 646-3115/3138. — ? ? ?

HAWAII: State Board of Registration for Professional Engineers, P. O. Box 3469, Honolulu 96801. Executive Secretary, Telephone: (808) 586-2702. — N N Y

IDAHO: Board of Professional Engineers, 600 S. Orchard, Suite A, Boise 83705-1242. Executive Secretary, Telephone: (208) 334-3860, dcurtis@ipels.state.id.us, www.state.id.us/ipels. — Y Y Y

ILLINOIS: State Board of Professional Engineers, 320 West Washington, 3rd Fl, Springfield 62786. Unit Manager, Telephone: (217) 785-0820, question@dpr084.1.state.il.us, www.state.il.us. — N Y N

INDIANA: State Board of Registration for Professional Engineers, 302 W. Washington St., E034, Indianapolis 46204. Executive Director, Telephone: (317) 232-3902, www.ai.org/pla. — Y Y 'Y'

IOWA: Engineering Examining Board, 1918 S.E. Hulsizer, Ankeny 50021. Executive Secretary, Tel: (515) 281-5602, jolene.schmitt@comm7,state.ia.us, www.state.ia.us/proflic. — N Y Y

KANSAS: State Board of Technical Professions, 900 Jackson, Suite 507, Topeka 66612. Executive Secretary, Telephone: (785) 296-3053, www.ink.org/public/ksbtp. — Y N Y

KENTUCKY: State Board of Licensure for Professional Engineers, 160 Democrat Dr., Frankfort 40601. Executive Director, Telephone: (502) 573-2680, larry.perkins@mail.state.ky.us, www.kyboels. — Y Y Y

NCEES Book?	Gen'l Exam?	Calculators?	
Y	Y	Y	**LOUISIANA:** State Board of Registration for Professional Engineers, 10500 Coursey Blvd Suite 107, Baton Rouge. Executive Secretary, Telephone: (225) 295-8522, www.lapels.com.
Y	Y	Y	**MAINE:** State Board of Registration for Professional Engineers, 92 State House, Station, Augusta 04333-0092. Secretary, Telephone: (207) 287-3236, pengineers@ctel.net, www.professionalsmaineusa,com.
N	Y	Y	**MARYLAND:** Board for Professional Engineers, 500 N. Calvert St, Rm 308, Baltimore 21202-3651. Executive Secretary, Telephone: (410) 230-6322, dmatricciani@dllr.state.md.us, www.dllr.state.md.us.
N	Y	Y	**MASSACHUSETTS:** State Board of Registration of Professional Engineers, 239 Canseway St, Boston 02114. Secretary, Telephone: (617) 727-3074, marie.e.deveau@state.ma.us, www.state.ma.us/reg.
'Y'	Y	Y	**MICHIGAN:** Board of Professional Engineers, P. O. Box 30018, Lansing 48909. Administrative Secretary, Telephone: (517) 241-9253, jack.sharpe@cis.state.mi.us. (NCEES Handbook provided upon signup if requested.)
Y	N	N	**MINNESOTA:** State Board of Registration for Engineers, 85 E. 7th Pl, Suite 160, St. Paul 55101. Executive Secretary, Telephone: (651) 296-2388, sheri.lindemann@state.mn.us, www.aelslagid.state.mn.us.
Y	Y	Y	**MISSISSIPPI:** State Board of Registration for Professional Engineers, P. O. Box 3, Jackson 39205. Executive Director, Telephone: (601) 359-6160, information@pepls.state.ms.us, www.peplsstate.ms.us.
N	Y	Y	**MISSOURI:** Board of Professional Engineers, P. O. Box 184, Jefferson City 65102. Executive Director, Telephone: (573) 751-0047, moapels@mail.state.mo.us, www.ecodev.state.mo.us/pr/apels.
N	Y	Y	**MONTANA:** State Board of Professional Engineers and Land Surveyors, Department of Commerce, 111 N. Jackson, P. O. Box 200513, Helena 59620-0513. Administrative Secretary, Telephone: (406) 444-1667, compolpel@state.my.us, www.com.state.mt.us/license/POL/pol_boards.
Y	N	Y	**NEBRASKA:** State Board of Professional Engineers, 301 Centennial Mall South, 6th Fl, Lincoln 68508. Executive Director, Telephone: (402) 471-2021, execdir@nol.org, www.nol.org/home/NBOP.
Y	Y	Y	**NEVADA:** State Board of Professional Engineers, 1755 East Plum Lane, Ste. 135, Reno 89502. Executive Secretary, Telephone: (775) 688-1231, nevengsur@natinfo.net, www.state.nv.us/BOE.

State Boards of Registration

	NCEES Book?	Gen'l Exam?	Calculators?

NEW HAMPSHIRE: State Board of Professional Engineers, 57 Regional Drive, Concord 03301. Executive Secretary, Telephone: (603) 271-2219, llavertu@nhsa.state.nh.us, www.state.nh.us/jtboard/home. **Y Y Y**

NEW JERSEY: State Board of Professional Engineers and Land Surveyors, P. O. Box 45015, Newark 07101. Executive Secretary-Director, Telephone: (973) 504-6460. **Y Y Y**

NEW MEXICO: State Board for Professional Engineers, 1010 Marquez Pl., Santa Fe 87501. Secretary, Telephone: (505) 827-7561, amanda.lopez@state.nm.us, www.state.nm.us/pepsboard. **Y Y N**

NEW YORK: State Board for Engineering, Cultural Education Center, Rm 3019, Albany 12230. Executive Secretary, Telephone: (518) 474-3846, enginbd@mail.nysed.gov, www.nysed.gov/prof/pe. **N Y Y**

NORTH CAROLINA: State Board of Professional Engineers, 310 W Millbrook Rd, Raleigh 27609. Executive Secretary, Telephone: (919) 841-4000, ncboard@ncbels.org, www.ncbels.org. **Y Y Y**

NORTH DAKOTA: State Board of Registration for Professional Engineers, P. O. Box 1357, Bismarck 58502. Executive Secretary, Telephone: (701) 258-0786. **Y Y 'N'**

OHIO: State Board of Registration for Professional Engineers, 77 S. High St., 16th Fl., Columbus 43266-0314. Executive Secretary, Telephone: (614) 466-3650, mjacob@mail.peps.state.oh.us, www.peps.state.oh.us. **Y Y Y**

OKLAHOMA: State Board of Registration for Professional Engineers, 201 N.E. 27th Street, Rm 120, Oklahoma City, 73105-2788. Executive Secretary, Telephone: (405) 521-2874, www.okpels.org. **Y Y Y**

OREGON: State Board of Engineering Examiners, Department of Commerce, 728 Hawthorne Ave NE, Salem 97301. Executive Secretary, Telephone: (503) 362-2666, grahame@osbeels.org, www.osbeels.org. **Y Y Y**

PENNSYLVANIA: State Registration Board for Professional Engineers, P. O. Box 2649, Harrisburg 17105-2649. Administrative Assistant, Telephone: (717) 783-7049, engineer@pados.dos.state.pa.us, www.dos.state.pa.us/bpoa/engbd. **N Y Y**

PUERTO RICO: Board of Examiners of Engineers, P. O. Box 9023271, San Juan 00907-3271. Director, Examining Boards, Telephone: (728) 722-4816. **? ? ?**

RHODE ISLAND: Board of Registration for Professional Engineers, 1 Capitol Hill, 3rd Fl, Providence 02908, Administrative Assistant, Ph: (401) 222-2565. **N N Y**

NCEES Book?	Gen'l Exam?	Calculators?	
Y	Y	Y	SOUTH CAROLINA: State Board of Registration for Professional Engineers, P. O. Box 11597, Columbia 29211-1597. Agency Director, Tele-phone: (803) 896-4422, engls@mail.llr.state.sc.us, www.llr.state.sc.us.
N	Y	Y	SOUTH DAKOTA: Board of Technical Professions, 2040 West Main St, Suite 304, Rapid City 57702-2447. Executive Secretary, Telephone: (605) 394-2510, snwhillpe@aol.com, www.state.sd.us/dcr/engineer.
N	Y	Y	TENNESSEE: State Board of Engineering Examiners, 500 James Robertson Pkwy, 3rd Fl,, Nashville 37243. Administrator, Telephone: (615) 741-3221, bbowling@mail.state.tn.us, www.state.tn.us/commerce/ae.
N	Y	Y	TEXAS: Board of Professional Engineers, P. O. Drawer 18329, Austin 78760-8329. Executive Director, Telephone: (512) 440-7723, peboard@mail.capnet.state.tx.us, www.main.org/peboard.
N	Y	Y	UTAH: Division of Occupational and Professional Licensing, P. O. Box 146741, Salt Lake City 84114-6741. Director, Telephone: (801) 530-6511, brcmrc.brdopl.dfairhur@email.state.ut.us.
Y	Y	Y	VERMONT: State Board of Registration for Professional Engineering, 26 Terrace St, Drawer 09, Montpelier 05609-1106. Executive Secretary, Telephone: (802) 828-2875, cpreston@sec.state.vt.us, www.sec.state.vt.us.
N	Y	Y	VIRGINIA: State Board of Professional Engineers, 3600 W Broad St, Richmond 23230. Assistant Director, Telephone: (804) 367-8512, apelsla@dpor.state.va.us, www.va.us/dpor.
N	Y	Y	VIRGIN ISLANDS: Board for Architects, Engineers and Land Surveyors, Bldg 1, Sub-Base, Rm 205, St. Thomas 00802. Secretary, Telephone: (340) 773-2226.
Y	Y	Y	WASHINGTON: State Board of Registration for Professional Engineers and Land Surveyors, P. O. Box 9649, Olympia 985047-9649 Executive Secretary, Telephone: (360) 753-6966, engineers@dol.wa.gov, www.wa.gov/dol/bpd/engfront.
N	Y	Y	WEST VIRGINIA: State Board of Registration for Professional Engineers, 608 Union Building, Charleston 25301-2703. Executive Director, Telephone: (304) 558-3554.
Y	Y	Y	WISCONSIN: State Examining Board of Professional Engineers, P. O. Box 8935, Madison 53708-8935. Administrator, Telephone: (608) 266-5511, dorl@mail.state.wi.us, www.badger.state.wi.us/agencies/drl.
Y	Y	Y	WYOMING: State Board of Examining Engineers, 2424 pioneer Ave, Suite 400, Cheyenne 82001. Secretary-Accountant, Telephone: (307) 777-6155, cturk@wyoming.com, www.wrds.uwyo.edu/wrds/borpe/borpe.

The FE/EIT Review for Civil Engineering

edited by Merle C. Potter

Chapters

1. Geotechnical .... Wolff
2. Structural Analysis .... Hatfield
3. Transportation .... Maleck
4. Airport Design .... McKelvey
5. Pavement Design .... Baladi
6. Environmental Engineering .... Masten
7. Hydrology .... Potter

Overview of the FE/Civil Topics

Most of the problems on the Civil Engineering test of the afternoon session can be worked using pages 87 to 98 of the NCEES Handbook, 3rd ed. Many of the problems, however, on the afternoon Civil Engineering test will come from the first part of the NCEES Handbook, material used for the General tests of the morning and afternoon sessions. Since all engineers must take the General morning exam, it is assumed that you are familiar with that material. You must realize, however, that you will have to make reference to equations, tables, and charts located in those first 83 pages of the NCEES Handbook when you are working problems in the afternoon Civil Engineering session. The General engineering sections that you should be particularly familiar with are the sections on Mathematics, Statics, Mechanics of Materials, and Fluid Mechanics. A review of General engineering material is found in our *Fundamentals of Engineering*, a review book written specifically for the FE/EIT exam.

What You Need to Know About the NCEES Handbook

Do I Have to Take the Discipline Test?

If, after you study this Civil Engineering review, and take the Civil Engineering Discipline Practice Exam, you feel that you could more easily pass the General exam of the afternoon session, you should register to take the afternoon General test. If you have already registered for the Civil Engineering afternoon test, you may contact your state board and request a change. Make sure your state board recognizes a *pass* on the General afternoon test for your particular discipline before you make that decision; the vast majority of the state boards will recognize a pass on any of the afternoon tests by any registrant.

Why This Review Is Focused on the NCEES Handbook

In order to provide a very efficient, yet effective review, this DS Review has been designed to review only the material associated with the formulas and figures presented in the NCEES Handbook. The Handbook is the only material allowed in the exam. Thus the Handbook makes an efficient base to study around. It is assumed that you have a copy of the NCEES Handbook, 3rd ed. If you do not, you can obtain one from Great Lakes Press by calling 1-800-837-0201.

GLP Offers a Concise Review for the A.M. Session & Other Study Aids

We also have an effective review for the General tests, both morning and afternoon, entitled *Fundamentals of Engineering*. It is used at hundreds of engineering schools across the nation.

For those who only need a quick refresher or readiness check for the General morning session, we offer *FE/EIT Quick Prep*.

We also distribute the HP 48GX calculator, which is pre-programmed with many of the equations you'll need when working problems on the FE/EIT exam. Also, we've developed *Jump Start the HP 48G/GX*, a 240-page book designed to efficiently help engineers get the most out of the HP 48G/GX. You can order these items by calling 1-800-837-0201.

1. Geotechnical Engineering

by Thomas F. Wolff

Geotechnical engineering may be subdivided into two broad areas of subject matter, each of which is often a separate undergraduate course. The first, *soil mechanics*, deals with the properties, classification, and stress-strain behavior (compressibility and strength) of soil materials and the movement of water through soils. The second, traditionally called *foundation engineering*, and more recently called *geotechnical design*, involves the application of the theory of soil mechanics to problems involving foundations, slopes and retaining structures.

The geotechnical "Definitions" cited in the NCEES Reference Handbook are in some cases definitions and in some cases equations. They relate to commonly encountered geotechnical engineering problems; however, many of the variables used in the definitions are themselves not defined. Furthermore, some are valid only for special cases and cannot always be used for general cases. Hence, this chapter provides a brief review to define all variables and describe limitations of the equations.

1.1 Weight–Mass–Volume Problems

Definitions and Unit Conversions

Weight–volume and mass–volume problems involve determining quantities, ratios or proportions of soil components working from other known quantities, ratios or proportions. In an actual volume of soil, the solids, water and (if not saturated) air are arranged in a complex fashion as shown in Figure 1.1. To simplify visualization of the problem, a *phase diagram* may be drawn. In a phase diagram, shown in Figure 1.2, the solid particles are shown as a single contiguous mass at the bottom of the diagram. Above the solids is shown the water, and above that, the air. On the left side of the diagram are shown the total volume V, the volume V_s of solids, the volume V_w of water, the volume V_a of air, and the volume V_v of voids, which is the combined volume of water and air. On the right side are shown the total weight W, the weight W_s of solids, and the weight W_w of water. The weight of air is taken as zero.

Figure 1.1 The Soil-Water System

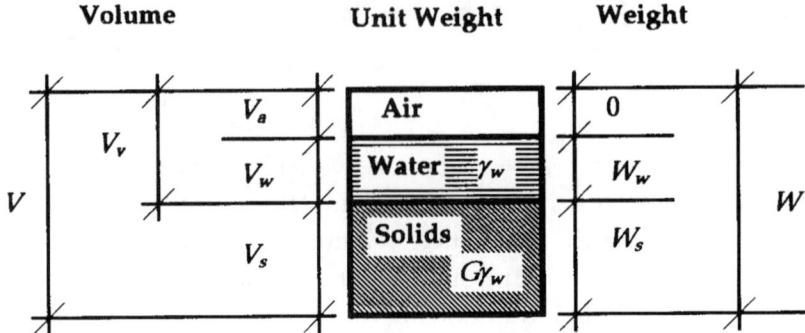

Figure 1.2 Phase Diagram

Observe that $V_v = V_w + V_a$, $V = V_s + V_v$, and $W = W_s + W_w$.

Relationships regarding soil and water volumes include the *void ratio e*, which is the ratio of the volume of voids to the volume of solids,

$$e = \frac{V_v}{V_s}$$

the *porosity η*, which is the ratio of the volume of voids to the total volume,

$$\eta = \frac{V_v}{V}$$

and the *degree of saturation S*, which is the ratio of the volume of water to the volume of voids and is usually expressed as a percentage,

$$S = \frac{V_w}{V_v} \times 100\%$$

The relationship involving soil and water weights is the *water content w* or *moisture content*, which is the ratio of the weight of water to the weight of solids,

$$w = \frac{W_w}{W_s} \times 100\%$$

Ratios of weights to volumes are termed *unit weights*. The *total unit weight γ* is the total weight of a soil mass (soil, water and air) divided by the total volume occupied, including the pore volume,

$$\gamma = \frac{W}{V}$$

The *saturated unit weight* γ_{sat} is the total unit weight obtained if the entire void volume were filled with water ($S = 100\%$ and $V_w = V_v$).

The *dry unit weight* γ_d is the ratio of the weight of solids to the total volume,

$$\gamma_d = \frac{W_s}{V}$$

The dry unit weight is to track the weight of soil solids in an entire occupied volume of solids and voids (e.g., the volume of a truck bed). For example,

$$W_s = \gamma_d V$$

The *specific gravity* G is the ratio of the unit weight of the soil solids to the unit weight of water,

$$G = \frac{\gamma_s}{\gamma_w}$$

hence, the *unit weight of solids* γ_s can be expressed as

$$\gamma_s = G\gamma_w = \frac{W_s}{V_s}$$

For most of the minerals comprising soil solids, G is in the range 2.65 to 2.75; $G = 2.70$ is a common assumption when tests results are not available.

Problems involving SI units may be phrased in the context of masses and mass densities rather than weights and unit weights; however, unit weights must ultimately be obtained for stress calculations. Mass density is denoted by ρ. As weight W equals mass m times the acceleration of gravity g ($W = mg$), the unit weight values defined above are in fact the product of mass densities and g,

$$\gamma = \rho g$$
$$\gamma_d = \rho_d g$$

In the SI system, mass densities are commonly expressed as Mg/m³, kg/m³, or g/mL,

$$1 \text{ Mg/m}^3 = 1000 \text{ kg/m}^3 = 1 \text{ g/mL}$$

Recall the mass density and unit weight of water,

Mass density of water: $\rho_w = 1 \text{ Mg/m}^3 = 1000 \text{ kg/m}^3 = 1 \text{ g/mL}$

Unit weight of water: $\gamma_w = 9.807 \text{ kN/m}^3 = 9807 \text{ N/m}^3 = 62.4 \text{ lb/ft}^3$

Example 1.1

Convert a total mass density of $\rho = 1900 \text{ kg/m}^3$ to a total unit weight for use in stress calculations.

(A) 19.0 kN/m³
(B) 18.63 kN/m³
(C) 1.90 kN/m³
(D) 1.863 kN/m³

Solution:

$$\gamma = \rho g = 1900 \text{ kg/m}^3 (9.807 \text{ m/s}^2) = 18\,630 \text{ kg/m}^2\text{s}^2$$
$$= 18\,630 \text{ N/m}^3 = 18.63 \text{ kN/m}^3. \text{ The answer is B.}$$

Example 1.2

Soil solids have a specific gravity of $G = 2.71$.

1. Find the unit weight of solids in SI units (kN/m^3).
   - (A) 26.6 kN/m³
   - (B) 27.1 kN/m³
   - (C) 169.1 kN/m³
   - (D) 2.71 kN/m³

2. Find the unit weight of solids in English units (lb/ft^3).
   - (A) 26.6 lb/ft³
   - (B) 2.71 lb/ft³
   - (C) 169.1 lb/ft³
   - (D) 62.4 lb/ft³

Solutions:

1. **A** $\gamma_s = G\gamma_w = 2.71(9.807 \text{ kN/m}^3) = 26.58 \text{ kN/m}^3$
2. **C** $\gamma_s = G\gamma_w = 2.71(62.4 \text{ lb/ft}^3) = 169.10 \text{ lb/ft}^3$

Example 1.3

A soil mass has a total unit weight of $\gamma = 130 \text{ lb/ft}^3$. Find the total unit weight in SI units (kN/m^3).
- (A) 13.0 kN/m³
- (B) 4.04 kN/m³
- (C) 2.08 kN/m³
- (D) 20.4 kN/m³

Solution:

Although a chain of conversion factors could be used, conversions can be simplified by taking the ratio of the unit weight of water in SI units to that in English units as an equality. (It's the same water.) Thus,

$$\gamma = 130 \text{ lb/ft}^3 \, [(9.807 \text{ kN/m}^3)/(62.4 \text{ lb/ft}^3)] = 20.43 \text{ kN/m}^3$$

The answer is D.

Solving Problems with Known Quantities

Although weight–volume problems can be worked directly with equations, errors are reduced and checks are provided if such problems are worked by drawing a phase diagram and performing the following steps.

- Fill in the known weights, volumes, and unit weights.
- Multiply known volumes by the corresponding unit weights to obtain weights. Divide known weights by the corresponding unit weight to obtain volumes. Use at least four significant digits to ensure that the results will balance when values are summed in the vertical direction and multiplied or divided in the horizontal direction.
- Where values of some relationships (γ, e, w, S, etc.) are given, use the definitions to obtain additional weights or volumes. Continue multiplication and division across the diagram and addition and subtraction up and down the sides until all weights and volumes are known and all quantities balance.
- Calculate the required values from the completed and checked diagram.

Example 1.4

A mold having a volume of 0.10 ft^3 was filled with moist soil. The weight of the soil in the mold was found to be 12.00 lb. The soil was oven dried and the weight after drying was 10.50 lb. The specific gravity of solids was known (or assumed) to be 2.70.

1. Find the water content.
   (A) 12.5%
   (B) 14.3%
   (C) 38.6%
   (D) 24.0%

2. Find the void ratio.
   (A) 0.604
   (B) 0.386
   (C) 0.143
   (D) 0.377

3. Find the porosity.
   (A) 0.604
   (B) 0.386
   (C) 0.143
   (D) 0.377

4. Find the degree of saturation.
   (A) 14.3%
   (B) 100%
   (C) 63.8%
   (D) 24.0%

5. Find the total unit weight.
   (A) 120 lb/ft³
   (B) 105 lb/ft³
   (C) 165 lb/ft³
   (D) 62.4 lb/ft³

6. Find the dry unit weight.
   (A) 120 lb/ft³
   (B) 105 lb/ft³
   (C) 165 lb/ft³
   (D) 62.4 lb/ft³

Solutions:

A phase diagram is drawn and the known quantities are entered (shown in bold). The remaining quantities are then calculated as follows:

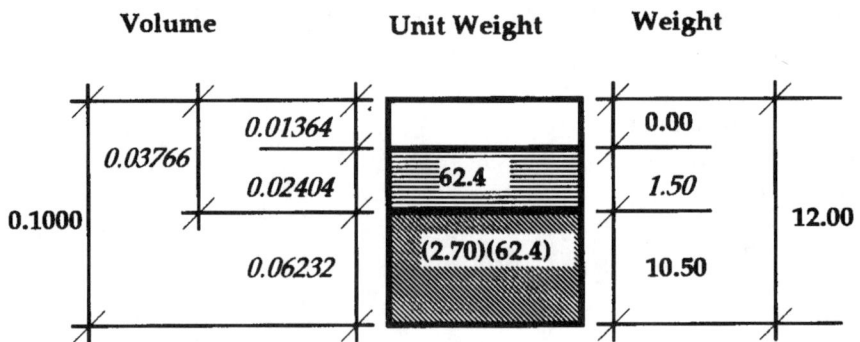

The weight and volume of water are, respectively,

$W_w = W - W_s = 12.00 - 10.50 = 1.50$ lb

$V_w = W_w / \gamma_w = 1.50 / 62.4 = 0.02404$ ft³

The volume of solids is

$$V_s = \frac{W_s}{G\gamma_w} = \frac{10.50}{2.70 \times 62.40} = 0.06232 \text{ ft}^3$$

The volume of air is

$V_a = V - V_w - V_s = 0.10000 - 0.02404 - 0.06232 = 0.01364$ ft³

The volume of voids is

$V_v = V_w + V_a = 0.02402 + 0.01364 = 0.03766$ ft³

The phase diagram is now complete and all the desired solutions can be obtained:

1. **B** $w = W_w/W_s = 1.50 / 10.50 = 0.143 = 14.30\%$
2. **A** $e = V_v/V_s = 0.03766 / 0.06232 = 0.604$

3. **D** $\eta = V_v/V = 0.03766 / 0.1000 = 0.377$
4. **C** $S = (V_w/V_v) 100\% = (0.02404 / 0.03768)(100\%) = 63.8\%$
5. **A** $\gamma = W/V = 12.00 / 0.100 = 120 \text{ lb/ft}^3$
6. **B** $\gamma_s = Ws/V = 10.50 / 0.100 = 105 \text{ lb/ft}^3$

Example 1.5

The soil in Example 1.4 becomes saturated by accumulating water. Assuming it retains the same total volume,

1. Find the saturated water content.
   (A) 19.6%
   (B) 37.6%
   (C) 22.4%
   (D) 60.4%

2. Find the saturated unit weight.
   (A) 128.5 lb/ft³
   (B) 105 lb/ft³
   (C) 120 lb/ft³
   (D) 165 lb/ft³

3. Find the dry unit weight.
   (A) 128.5 lb/ft³
   (B) 105 lb/ft³
   (C) 165 lb/ft³
   (D) 120 lb/ft³

Solutions:

After saturation, $V_w = 0.03768 \text{ ft}^3$. Taking $W_w = V_w \gamma_w$ the new weight of water is $(0.03768)(62.4) = 2.350$ lb. The phase diagram is modified as shown and solutions follow.

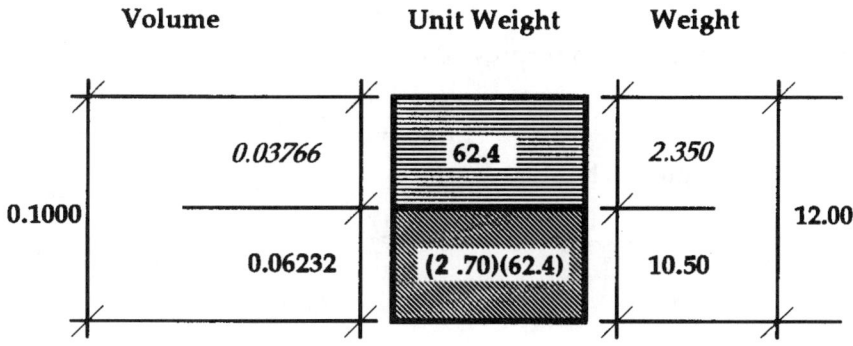

1. **C** $\quad w = \dfrac{W_w}{W_s} = \dfrac{2.350}{10.50} = 0.2239$ or 22.4%

2. **A** $\quad \gamma = \dfrac{W}{V} = \dfrac{10.50 + 2.35}{0.100} = 128.5 \text{ lb/ft}^3$

3. **B** $\quad \gamma_d = \dfrac{W_s}{V} = \dfrac{10.50}{0.100} = 105.0 \text{ lb/ft}^3$

Note that the dry unit weight does not change with an addition of water at total constant volume.

Solving Problems Involving Only Relationships

For problems where only relationships (w, e, S, etc.) are given, only relationships may be calculated. The phase diagram approach may still be used by assuming a value for one weight or volume. While any quantity can be assumed, certain assumptions greatly simplify calculations:

- If w is given, assume $W_s = 1.00$ or 100.00 (lb or kg); then $W_w = w$ or $100w$.
- If e is given, assume $V_s = 1.000 \text{ ft}^3$ or m^3; then $V_v = e$.
- If γ is given, assume $V = 1.000 \text{ ft}^3$ or m^3; then $W = \gamma$.
- If γ_d is given, assume $V = 1.000 \text{ ft}^3$ or m^3; then $W_s = \gamma_d$.

It is recommended that weight-volume problems be solved using phase diagrams rather than just formulas; nevertheless, a very useful equation that relates four different relationships is

$$Se = wG$$

For saturated soils ($S = 100\%$) this becomes

$$e = wG$$

The relationships between the void ratio and porosity are

$$e = \dfrac{\eta}{1-\eta}$$

and

$$\eta = \dfrac{e}{1+e}$$

The total unit weight can be expressed as:

$$\gamma = \dfrac{(G + Se) \times \gamma_w}{1+e}$$

The dry unit weight can be obtained as:

$$\gamma_d = \dfrac{G\gamma_w}{1+e}$$

Example 1.6

A soil has a water content of 20 percent, a void ratio of 0.800, and a specific gravity of 2.65.

1. Find the degree of saturation.
   (A) 80.0%
   (B) 20.0%
   (C) 66.3%
   (D) 53.0%

2. Find the porosity.
   (A) 0.444
   (B) 0.800
   (C) 0.530
   (D) 0.500

3. Find the total unit weight.
   (A) 165.4 lb/ft³
   (B) 198.4 lb/ft³
   (C) 110.2 lb/ft³
   (D) 91.9 lb/ft³

4. Find the dry unit weight.
   (A) 165.4 lb/ft³
   (B) 91.9 lb/ft³
   (C) 110.2 lb/ft³
   (D) 198.4 lb/ft³

Solutions:

Assume that the volume of solids $V_s = 1.000$ ft³; the phase diagram is drawn as follows.

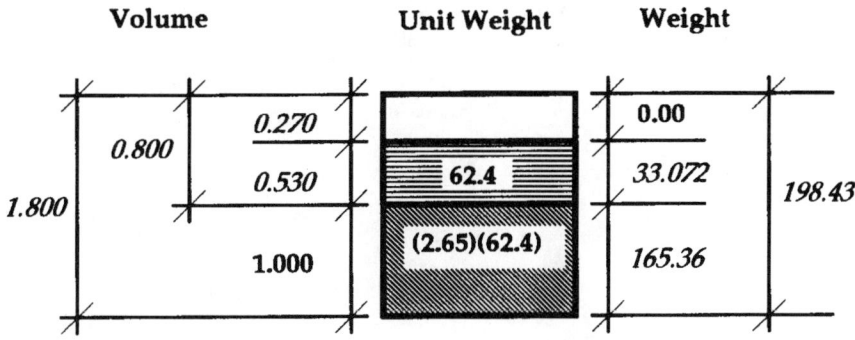

Remaining quantities are calculated as:

$V_v = eV_s = (0.800)(1.000) = 0.800 \text{ ft}^3$

$W_s = V_s G \gamma_w = (1.000)(2.65)(62.4) = 165.36 \text{ lb}$

$W_w = wW_s = (0.20)(165.36) = 33.072 \text{ lb}$

$V_w = W_w/\gamma_w = (33.072)/(62.4) = 0.530 \text{ ft}^3$

$V_a = V_v - V_w = 0.800 - 0.530 = 0.270 \text{ ft}^3$

At this point, the weights and volumes of all components are known for the assumed 1.000 ft³ of solids and the required solutions can be calculated:

1. **C** $\quad S = \dfrac{V_w}{V_v} \times 100\% = \dfrac{0.530}{0.800} \times 100\% = 66.3\%$

   Alternatively,

   $$S = \dfrac{wG}{e} = \dfrac{(.20)(2.65)}{0.800} = 0.6625 \text{ or } 66.3\%$$

2. **A** $\quad \eta = \dfrac{V_v}{V} = \dfrac{0.800}{1.000 + 0.800} = 0.444$

   Alternatively,

   $$\eta = \dfrac{e}{1+e} = \dfrac{0.800}{1.800} = 0.444$$

3. **C** $\quad \gamma = \dfrac{165.36 + 33.072}{1.800} = 110.2 \text{ lb/ft}^3$

   Alternatively,

   $$\gamma = \dfrac{(G + Se)\gamma_w}{1+e}$$

   $$= \dfrac{(2.65 + (0.6625 \times 0.800))62.4}{1.800} = 110.2 \text{ lb/ft}^3$$

4. **B** $\quad \gamma_d = \dfrac{165.36}{1.800} = 91.9 \text{ lb/ft}^3$

   Alternatively,

   $$\gamma_d = \dfrac{G\gamma_w}{1+e} = \dfrac{(2.65)(62.4)}{1.800} = 91.9 \text{ lb/ft}^3$$

1.2 Relative Density

The density of sands and gravels is often specified and measured in terms of the *relative density* D_d (the notation in the NCEES Handbook; note that many texts use D_r or I_D). The term is a misnomer, as it deals with void ratio, not density. It scales the actual void ratio as a fraction of the range between the maximum and minimum void ratios:

$$D_d = \frac{e_{max} - e}{e_{max} - e_{min}} \times 100\%$$

To determine the relative density using actual, minimum, and maximum dry unit weight values rather than void ratios, the above equation can be manipulated to obtain:

$$D_d = \frac{1/\gamma_{min} - 1/\gamma_d}{1/\gamma_{min} - 1/\gamma_{max}} \times 100\%$$

Example 1.7

Tests are made to determine the maximum and minimum void ratios for a sand, and it is found that $e_{min} = 0.500$ and $e_{max} = 0.700$. A well-compacted sample in the field is found to have a void ratio of $e = 0.550$. What is the relative density?

(A) 25%
(B) 55%
(C) 100%
(D) 75%

Solution:

$$D_d = \frac{e_{max} - e}{e_{max} - e_{min}} \times 100\% = \frac{0.70 - 0.55}{0.70 - 0.50} \times 100\% = 75\%$$

The answer is **D**.

Example 1.8

Laboratory tests on a sand indicate a minimum dry unit weight of 16.0 kN/m³ and a maximum dry unit weight of 18.0 kN/m³. It is specified to be compacted to at least 60% relative density. What is the minimum acceptable unit weight that will pass the specifications?

(A) 17.20 kN/m³
(B) 17.14 kN/m³
(C) 10.80 kN/m³
(D) 9.60 kN/m³

Solution:

Substituting in $D_d = \dfrac{1/\gamma_{min} - 1/\gamma_d}{1/\gamma_{min} - 1/\gamma_{max}} \times 100\%$,

$$60\% = \dfrac{0.06250 - 1/\gamma_d}{0.06250 - 0.05556} \times 100\%$$

$$1/\gamma_d = 0.0625 - 0.600 \times 0.00694 = 0.05834$$

$$\gamma_d = 17.14 \text{ kN}/\text{m}^3$$

Note that four or more significant digits must be retained to preserve accuracy, as the problem involves a ratio of differences of inverses, a very numerically sensitive problem. The answer is **B**.

1.3 Grain-Size Characteristics of Soils

The *grain-size distribution* of sands and gravels is determined by a *sieve analysis* (or *mechanical analysis*). Opening sizes for commonly used standard sieves are shown in Table 1.1. A soil sample is passed over the sieves and the sieves are shaken until all particles have passed down to the sieve that retains them. The cumulative weight of all material larger than each opening size is divided by the total sample weight to obtain the *percentage retained* or the *percent coarser*; the cumulative weight of all material smaller than that size divided by the total sample weight is the *percentage passing* or the *percent finer*. The results are plotted as the percent finer (linear scale) versus the sieve opening or grain-size (log scale). The plot is referred to as a *grain size curve,* examples of which are presented in Figure 1.3.

Table 1.1 Opening Sizes of Standard Sieves

	Sieve Size		Opening	
	1.5	in	37.5	mm
	1		25	
	0.75		19	
	0.5		12.5	
No.:	4		4.75	mm
	10		2.00	
	20		0.850	
	40		0.425	
	70		0.212	
	100		0.150	
	200		0.075	

Certain sizes of interest may be read from the grain size curve. The D_{10} size, sometimes called the *effective size*, is the grain diameter for which 10 percent of the sample (by weight) is finer. The D_{50} size, called the *median grain size*, is the grain diameter for which 50 percent of the sample (by weight) is finer. In general, the notation D_{xx} refers to the grain diameter for which *xx* percent of the sample is finer by weight.

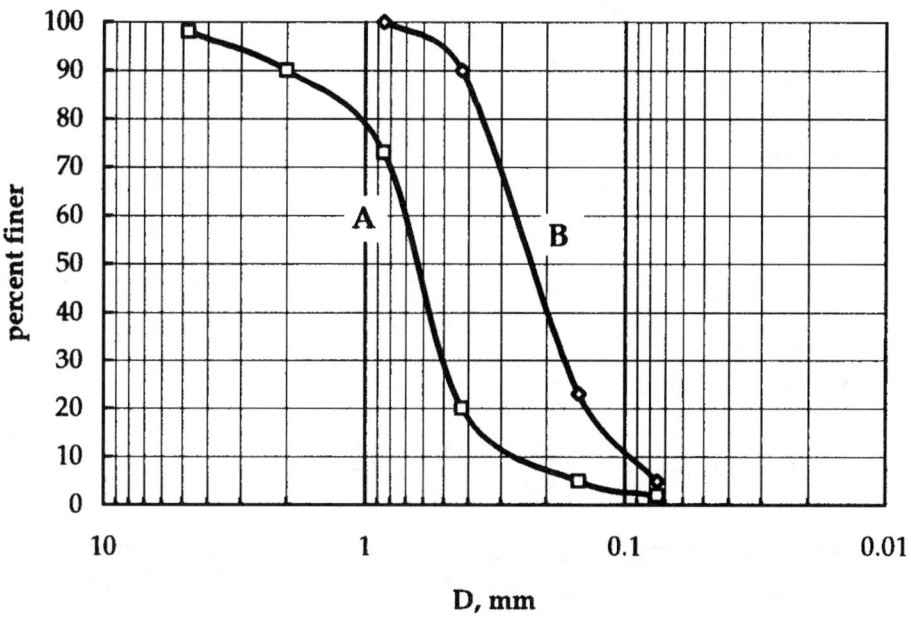

Figure 1.3 Grain Size Curves

Two parameters are used to describe the shape of the grain-size curve. The *coefficient of uniformity* c_u is:

$$c_u = \frac{D_{60}}{D_{10}}$$

It provides a measure of the slope of the curve. The *coefficient of curvature* or *coefficient of gradation* c_c is:

$$c_c = \frac{D_{30}^2}{D_{60}D_{10}}$$

It provides a measure of the smoothness of the curve.

Example 1.9

For the grain size curve A in Figure 1.3,

1. Determine the effective grain size, D_{10}.
   (A) 0.12 mm
   (B) 0.41 mm
   (C) 10.0 mm
   (D) 0.28 mm

2. Determine the coefficient of uniformity.
   (A) 2.5
   (B) 0.40
   (C) 0.70
   (D) 25

3. Determine the coefficient of curvature.
   (A) 2.5
   (B) 0.75
   (C) 1.33
   (D) 0.51

Solutions:

1. **D** Start at the 10% passing point on the vertical axis, move horizontally across to Curve A, then down to read $D_{10} = 0.28$ mm

2. **A** $D_{60} = 0.70$ mm, $c_u = \dfrac{D_{60}}{D_{10}} = \dfrac{0.70}{0.28} = 2.5$

3. **C** $D_{30} = 0.51$ mm, $c_c = \dfrac{D_{30}^2}{D_{60}D_{10}} = \dfrac{0.51^2}{(0.70)(0.28)} = 1.33$

Example 1.10

Refer to the grain size curve B in Figure 1.3.

1. Determine the effective grain size, D_{10}.
   (A) 0.019 mm
   (B) 0.27 mm
   (C) 1 mm
   (D) 0.095 mm

2. Determine the coefficient of uniformity.
   (A) 2.50
   (B) 2.84
   (C) 0.35
   (D) 0.27

3. Determine the coefficient of curvature.
   (A) 1.13
   (B) 0.17
   (C) 0.88
   (D) 5.88

Solutions:

1. **D** Start at the 10% passing point on the vertical axis, move horizontally across to Curve B, then down to read $D_{10} = 0.095$ mm

2. **B** $D_{60} = 0.27$ mm, $c_u = \dfrac{D_{60}}{D_{10}} = \dfrac{0.27}{0.095} = 2.84$

3. **A** $D_{30} = 0.17$ mm, $c_c = \dfrac{D_{30}^2}{D_{60}D_{10}} = \dfrac{0.17^2}{(0.27)(0.095)} = 1.13$

1.4 Atterberg Limits and Plasticity

Atterberg limits are specific water contents at which soil behavior changes. The upper limit of plasticity, above which soil behaves as a liquid, and below which soil behaves as a plastic solid, is the *liquid limit*, denoted LL (or w_L). The lower limit of plastic behavior, above which the soil behaves as a plastic solid, and below which the soil behaves as a brittle solid, is the *plastic limit*, denoted PL (or w_P). The water content below which the soil no longer reduces in volume with a reduction in water content is the shrinkage limit SL or w_S. Although Atterberg limits are water contents (decimals or percentages), they are by convention expressed as integers values (implying a percentage).

The *plasticity index*, denoted PI (or I_P) is the difference of the liquid limit and the plastic limit:

$$PI = LL - PL$$

The *liquidity index*, denoted LI, is a measure of the natural water content relative to the plastic limit and the liquid limit:

$$LI = \frac{w - PL}{LL - PL}$$

The shrinkage index, denoted SI, is the difference of the plastic limit and the shrinkage limit:

$$SI = PL - SL$$

Example 1.11

A soil has a natural water content of 40%. Its liquid limit is 50, the plastic limit is 30 and the shrinkage limit is 20.

1. Determine the plasticity index.
   - (A) 20
   - (B) 30
   - (C) 10
   - (D) 80

2. Determine the liquidity index.
   - (A) 2.00
   - (B) 0.75
   - (C) 0.50
   - (D) 0.30

3. Determine the shrinkage index.
   - (A) 20
   - (B) 30
   - (C) 80
   - (D) 10

Solutions:

1. **A** $PI = LL - PL = 50 - 30 = 20$

2. **C** $LI = \dfrac{w - PL}{LL - PL} = \dfrac{40 - 30}{50 - 30} = 0.50$

3. **D** $SI = PL - SL = 30 - 20 = 10$

1.5 Stresses in Soil

Normal stresses (σ) in a soil mass are carried partly by the soil skeleton as effective stress (σ') and partly by pore water pressure (u). Shear stresses (τ) must be carried entirely by the soil skeleton, since water cannot sustain shear stress. Terzaghi's *effective stress principle* states that:

$$\sigma = \sigma' + u$$

The vertical normal stress σ'_v on a horizontal plane may be calculated by dividing the overlying total weight of the soil by the area over which the soil acts:

$$\sigma = \frac{W}{A}$$

The NCEES Handbook uses the general equation $\sigma = P/A$. As the weight of soil is the unit weight times the volume, or $W = \gamma h_s A$, the above equation can be written

$$\sigma = \Sigma \gamma_s h_s$$

where the summation sign denotes that the unit weight and height of each soil layer must be treated separately.

The pore pressure, in conditions of a static water table, is the unit weight of water times the height of water above the plane on which stresses are calculated:

$$u = \gamma_w h_w$$

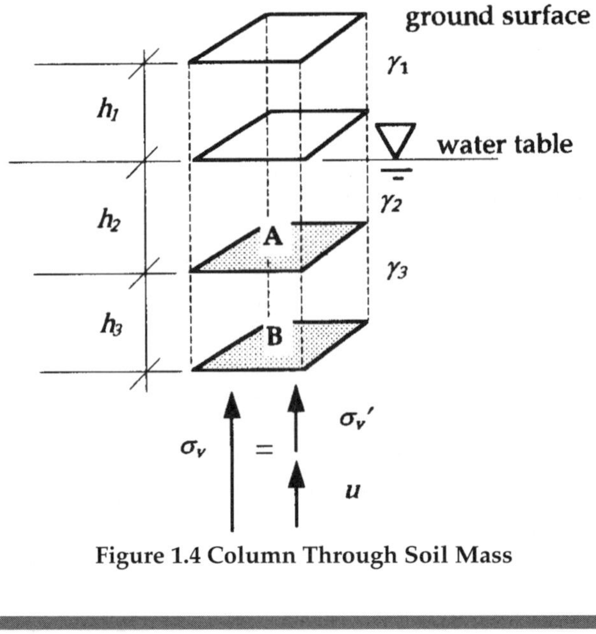

Figure 1.4 Column Through Soil Mass

Example 1.12

In Figure 1.4, assume $h_1 = 3$ ft, $\gamma_1 = \gamma_{moist} = 110$ lb/ft^3, $h_2 = 4$ ft, $\gamma_2 = \gamma_{sat} = 120$ lb/ft^3, $h_3 = 3$ ft, $\gamma_3 = \gamma_{sat} = 125$ lb/ft^3.

1. Determine the total vertical stress on plane A.
   - (A) 770 lb/ft^2
   - (B) 810 lb/ft^2
   - (C) 840 lb/ft^2
   - (D) 230 lb/ft^2

2. Determine the pore pressure on plane A.
   - (A) 437 lb/ft^2
   - (B) 187 lb/ft^2
   - (C) 250 lb/ft^2
   - (D) 62.4 lb/ft^2

3. Determine the effective vertical stress on plane A.
   - (A) 560 lb/ft^2
   - (B) 373 lb/ft^2
   - (C) 623 lb/ft^2
   - (D) 810 lb/ft^2

4. Determine the total vertical stress on plane B.
   - (A) 1250 lb/ft^2
   - (B) 1100 lb/ft^2
   - (C) 1200 lb/ft^2
   - (D) 1185 lb/ft^2

5. Determine the pore pressure on plane B.
   - (A) 624 lb/ft^2
   - (B) 437 lb/ft^2
   - (C) 62.4 lb/ft^2
   - (D) 187 lb/ft^2

6. Determine the effective vertical stress on plane B.
   - (A) 561 lb/ft^2
   - (B) 1185 lb/ft^2
   - (C) 748 lb/ft^2
   - (D) 626 lb/ft^2

Solutions:

1. **B** $\sigma_v = \Sigma \gamma_s h_s = (110)(3) + (120)(4) = 810 \text{ lb/ft}^2$

2. **C**   The water table is 4 ft above the plane, hence, $u = \gamma_w h_w = (62.4)(4.0) = 249.6 \text{ lb/ft}^2$

3. **A** $\sigma'_v = \sigma_v - u = 810 - 249.6 = 560.4 \text{ lb/ft}^2$

4. **D** $\sigma_v = \Sigma \gamma_s h_s = (110)(3) + (120)(4) + (125)(3) = 1185 \text{ lb/ft}^2$

5. **B** $u = \gamma_w h_w = (62.4)(4.0 + 3.0) = 436.8 \text{ lb/ft}^2$

6. **C** $\sigma'_v = \sigma_v - u = 1185 - 436.8 = 748.2 \text{ lb/ft}^2$

Example 1.13

In Figure 1.4, assume $h_1 = 2$ m, $\gamma_1 = \gamma_{moist} = 18$ kN/m³, $h_2 = 3$ m, $\gamma_2 = \gamma_{sat} = 19$ kN/m³, $h_3 = 2$ m, $\gamma_3 = \gamma_{sat} = 20$ kN/m³.

1. Determine the total vertical stress on plane A.
   - (A) 95 kN/m²
   - (B) 90 kN/m²
   - (C) 93 kN/m²
   - (D) 18 kN/m²

2. Determine the pore pressure on plane A.
   - (A) 29.4 kN/m²
   - (B) 49.1 kN/m²
   - (C) 9.8 kN/m²
   - (D) 93.0 kN/m²

3. Determine the effective vertical stress on plane A.
   - (A) 93.0 kN/m²
   - (B) 43.9 kN/m²
   - (C) 0 kN/m²
   - (D) 63.6 kN/m²

4. Determine the total vertical stress on plane B.
   - (A) 140 kN/m²
   - (B) 133 kN/m²
   - (C) 126 kN/m²
   - (D) 20 kN/m²

5. Determine the pore pressure on plane B.
   (A) 68.7 kN/m²
   (B) 19.6 kN/m²
   (C) 49.1 kN/m²
   (D) 9.8 kN/m²

6. Determine the effective vertical stress on plane B.
   (A) 83.9 kN/m²
   (B) 64.3 kN/m²
   (C) 113.4 kN/m²
   (D) 71.3 kN/m²

Solutions:

1. **C** $\sigma_v = \Sigma \gamma_s h_s = (18)(2) + (19)(3) = 93.0 \text{ kN/m}^2$ (or 93.0 kPa)

2. **A**   The water table is 3 m above the plane, hence,
$$u = \gamma_w h_w = (9.81)(3.0) = 29.4 \text{ kN/m}^2$$

3. **D** $\sigma'_v = \sigma_v - u = 93.0 - 29.4 = 63.6 \text{ kN/m}^2$

4. **B** $\sigma_v = \Sigma \gamma_s h_s = (18)(2) + (19)(3) + (20)(2) = 133 \text{ kN/m}^2$

5. **C** $u = \gamma_w h_w = (9.81)(3.0 + 2.0) = 49.1 \text{ kN/m}^2$

6. **A** $\sigma'_v = \sigma_v - u = 133.0 - 49.1 = 83.9 \text{ kN/m}^2$

1.6 Water Flow in Soil

Flow through Uniform Sections

The flow rate Q through a soil mass is given by *Darcy's Law*:
$$Q = vA = kiA$$
where Q is the flow (volume/time), v is the discharge velocity, k is the *coefficient of permeability* (length/time), i is the *hydraulic gradient*, and A is the cross-sectional area. The hydraulic gradient is the total head loss divided by the length of flow path over which that head loss occurs. Rearranging, the coefficient of permeability can be measured by creating a known gradient through a sample of known cross-sectional area, and measuring the flow:
$$k = \frac{Q}{iA}$$

Example 1.14

The tube shown is circular with a uniform diameter D and contains a soil sample of length L. A constant head loss H is maintained across the sample by pouring in water at point A, and permitting water to overflow at point B.

Given: $D = 2$ cm, $H = 10$ cm, $L = 20$ cm, and $k = 0.1$ cm/s.

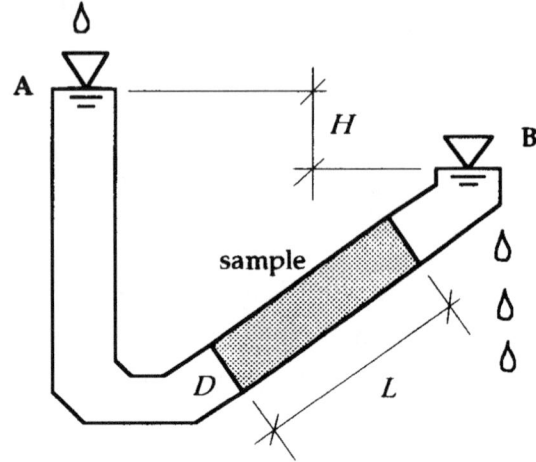

1. What is the cross-sectional area of the tube?
   (A) 12.57 cm²
   (B) 3.14 cm²
   (C) 4.0 cm²
   (D) 2.0 cm²

2. What is the gradient?
   (A) 0.5
   (B) 1.0
   (C) 10.0
   (D) 2.0

3. What is the discharge velocity?
   (A) 0.1 cm/s
   (B) 20 cm/s
   (C) 0.25 cm/s
   (D) 0.05 cm/s

4. What is the flow rate?
   (A) 0.629 cm³/s
   (B) 0.314 cm³/s
   (C) 0.157 cm³/s
   (D) 0.079 cm³/s

5. How long will it take to pass 1 liter of water through the sample?
    (A) 0.157 s
    (B) 1.77 hr
    (C) 1.77 min
    (D) 1.77 day

Solutions:

1. **B** $A = \pi D^2/4 = (3.14)(2)^2/4 = 3.14 \text{ cm}^2$
2. **A** $i = H/L = 10/20 = 0.5$
3. **D** $v = ki = (0.1)(0.5) = 0.05 \text{ cm/s}$
4. **C** $Q = kiA = (0.1)(0.5)(3.14) = 0.157 \text{ cm}^3/\text{s}$
5. **B** $t = (1000 \text{ cm}^3) / (0.157 \text{ cm}^3/\text{s}) = 6369 \text{ sec} \approx 1.77 \text{ hr}$

Example 1.15

Refer to the figure used in the previous problem. Given: $D = 5$ cm, $L = 20$ cm, $H = 40$ cm, and $Q = 1.96 \text{ cm}^3/\text{sec}$.

1. What is the cross-sectional area of the tube?
    (A) 78.54 cm²
    (B) 25 cm²
    (C) 19.64 cm²
    (D) 5 cm²

2. What is the gradient?
    (A) 0.5
    (B) 2
    (C) 1.0
    (D) 4.0

3. What is the coefficient of permeability?
    (A) 0.05 cm/s
    (B) 0.10 cm/s
    (C) 0.025 cm/s
    (D) 1.96 cm/s

4. What is the discharge velocity?
    (A) 0.2 cm/s
    (B) 0.1 cm/s
    (C) 0.4 cm/s
    (D) 1.96 cm/s

Solutions:

1. **C** $A = \pi D^2/4 = (3.14)(5)^2/4 = 19.64 \text{ cm}^2$
2. **B** $i = H/L = 40/20 = 2.0$
3. **A** $k = Q/(iA) = 1.96/[(2)(19.64)] = 0.05 \text{ cm/s}$
4. **B** $v = ki = (Q/iA)(i) = Q/A = 1.96/19.64 = 0.1 \text{ cm/s}$

Flow Through Non-Uniform Sections

For nonuniform flow regimes, flows and gradients can be determined from a *flow net*, a graphical solution of Laplace's equation obtained by trial-and-error sketching or computer solution. A flow net consists of a family of *equipotential lines*, which are contours of equal total head, and *flow lines*, which are curves describing selected flow paths of water particles. If a flow net is constructed for a homogeneous, isotropic flow regime such that the equipotential lines and flow lines intersect at right angles and form curvilinear squares, the flow through a unit depth (into the paper) cross-section can be expressed as:

$$Q = kH \frac{N_f}{N_e}$$

where H is the total head loss across the section, N_f is the number of flow channels, and N_e is the number of equipotential drops.

Example 1.16

A sheetpile is driven in a shallow lake and the area to one side is pumped down as shown. The foundation is sand underlain by impervious rock. A flow net is drawn as shown. $H = 5$ ft and $k = 0.2$ ft/min.

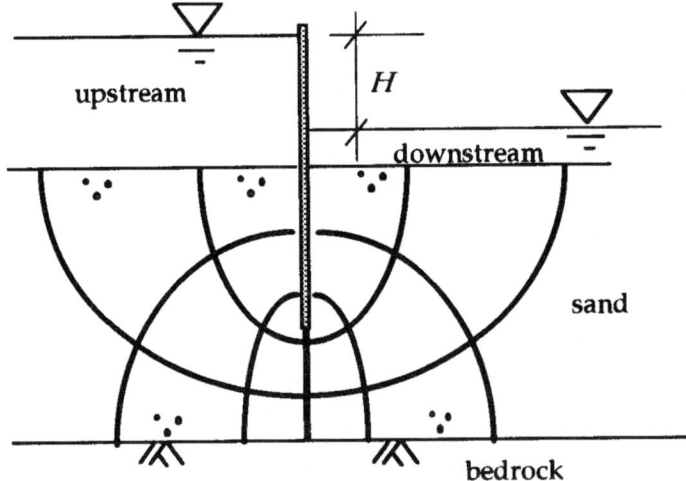

1. What is the flow per lineal foot of sheetpile?
   (A) 0.5 ft³/min
   (B) 1.0 ft³min
   (C) 2.0 ft³/min
   (D) 1.5 ft³/min

2. If the sheetpile and the excavation are 50 ft long (perpendicular to the paper), what capacity of pump is necessary to handle the flow?
   (A) 25 ft³/min
   (B) 5 ft³/min
   (C) 50 ft³/min
   (D) 100 ft³/min

3. If the sheetpile is extended upward to permit doubling H, what happens to the total flow?
   (A) it drops by half
   (B) it doubles
   (C) it stays the same
   (D) it increases by 41%

4. If the sheetpile is extended downward to cut through more of the sand, what happens to the flow?
   (A) it increases
   (B) it decreases
   (C) it stays the same
   (D) solution cannot be determined

Solutions:

The number of flow channels N_f is three: one is between the sheet pile and the upper drawn flow line, the second is between the two drawn flow lines, and the third is between the lower drawn flow line and the bedrock. As water flows along any flow line from upstream to downstream, it traverses 6 squares or equipotential drops; $N_e = 6$.

1. **A** $Q = kH(N_f/N_e) = (0.2)(5)(3/6) = 0.5$ ft²/min $= 0.5$ ft³/min/ft of length.

2. **A** $Q_{total} = Q \times$ length $= (0.5$ ft³/min/ft$)(50$ ft$) = 25$ ft³/min.

3. **B**   As the flow equation is linear in H, doubling H doubles the flow.

4. **B**   For any constant number of flow lines N_f, deepening the sheetpile increases the resistance to flow, increasing N_e and decreasing the flow.

1.7 Settlement of Saturated Clay

Void Ratio — Pressure Relationship

Increasing the effective stress on a soil causes a volume reduction as soil solids are pushed closer together and water is expelled from the pores. For low-permeability soils such as clay, the volume change is time-dependent and the process is termed *consolidation*. The parameters used to calculate the rate and amount of consolidation can be measured in a *consolidation test*. A cylindrical sample of soil is confined in a rigid ring and submerged in water with a porous stone above and below the soil. A load is applied and the reduction in soil height is measured as a function of time. When compression has essentially stopped, another load is applied and the process is repeated.

The test results are analyzed by plotting the final void ratio e under each load on a linear scale versus the applied pressure p (actually an effective vertical stress σ'_v) plotted on a log scale. The plot, often termed an *e-log p curve*, is shown in Figure 1.5.

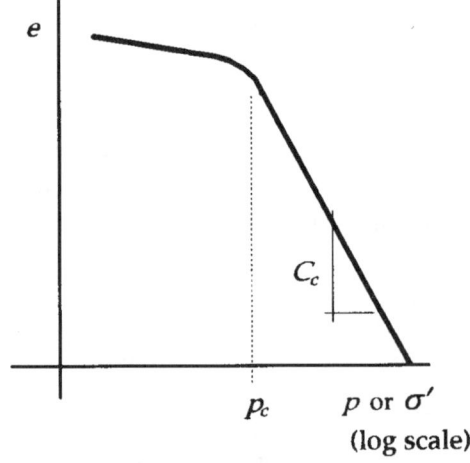

Figure 1.5 e - log p Consolidation Curve

The *preconsolidation pressure* p_c, or maximum effective stress the soil has been subjected to in the past, corresponds to the break in the curve. At pressures below the preconsolidation pressure, the curve has a relatively flat slope, and is termed the *recompression curve*. The soil in this stress range is *overconsolidated*. Above the preconsolidation pressure, the curve is steeper and is referred to as the *virgin consolidation curve*. In this range, the soil is said to be *normally consolidated*.

The virgin curve can be approximated as a straight line with a slope of C_c, termed the *compression index*. The compression index is

$$C_c = \frac{\Delta e}{\Delta \log_{10} p} = \frac{e_2 - e_1}{\log_{10} p_2 - \log_{10} p_1}$$

Although the compression index is always negative, the sign is not usually shown. The *recompression index* C_r can be calculated in the same manner using two points on the recompression curve.

The compression index of normally consolidated clays can be estimated from
$$C_c = 0.009(LL - 10)$$
where LL is the liquid limit expressed as a percentage.

Example 1.17

A normally consolidated clay layer has a liquid limit of 45. The effective stress at the midpoint of the layer is 2000 lb/ft².

1. Estimate the compression index.
   (A) 0.405
   (B) 0.004
   (C) 0.315
   (D) 0.003

2. Placement of a sand fill on the surface over a wide area causes a vertical stress increase of 800 lb/ft². How much will the void ratio at the midpoint of the layer decrease due to the stress increase?
   (A) 0.046
   (B) 0.125
   (C) 1.160
   (D) 0.315

Solutions:

1. **C** $C_c = 0.009(45-10) = 0.315$
2. **A** $\Delta e = C_c (\Delta \log_{10} p) = 0.315 [(\log_{10} 2800 - \log_{10} 2000)]$
   $= 0.315 \log_{10} (2800/2000) = 0.046$

Example 1.18

The pressure is doubled on a soil for which $C_c = 0.30$ and the initial void ratio e_i is 0.800 (note: the NCEES Handbook uses e_i; most texts use e_o). What is the void ratio after the soil consolidates under the pressure increase?
   (A) 0.090
   (B) 0.890
   (C) 0.830
   (D) 0.710

Solution:

1. **D** $\Delta e = C_c (\Delta \log_{10} p) = 0.30 (\log_{10} 2p - \log_{10} p) = 0.30 \times \log_{10} 2 = 0.090$
   $e_{final} = e_i - \Delta e = 0.800 - 0.090 = 0.710$

Magnitude of Settlement

The settlement of a clay layer can be calculated by determining the vertical strain ε_v as a function of depth and integrating the strain with respect to depth. As shown in Figure 1.6, if consolidation causes a void ratio reduction of Δe, the vertical strain ε_v is

$$\varepsilon_v = \frac{\Delta e}{1+e_i}$$

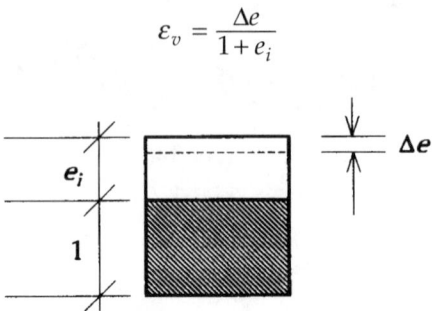

Figure 1.6 Void Ratio Reduction due to Consolidation

For a sufficiently thin layer of thickness H, the strain may be nearly constant, and the settlement ΔH can be calculated as the strain times the layer thickness:

$$\Delta H = H\varepsilon_v$$

Substituting in the previous equations for Δe, the following expression, found in the NCEES Handbook, is obtained:

$$\Delta H = H \frac{C_c}{1+e_i} \log_{10} \frac{p_i + \Delta p}{p_i}$$

Note that this equation is accurate only when

- the soil is normally consolidated (consolidation occurs along the virgin curve, with constant slope C_c)
- the layer thickness H is sufficiently small such that the initial pressure p_i and the pressure change Δp are good averages over the layer

Example 1.19

A 5 ft thick normally consolidated clay layer is located between two sand layers. Assume that $e_i = 0.900$ and $C_c = 0.35$. A surface load increases the existing average effective stress in the layer from 1000 lb/ft² by an additional 500 lb/ft².

1. Calculate the strain in the clay layer.
   (A) 0.0620
   (B) 0.0326
   (C) 0.1630
   (D) 0.3500

2. Calculate the settlement of the clay layer.
   (A) 1.96 in
   (B) 0.196 in
   (C) 0.033 in
   (D) 0.263 in

Solutions:

1. **B** $\Delta e = C_c (\Delta \log_{10} p) = 0.35 (\log_{10} 1500 - \log_{10} 1000)$

   $= 0.35 \log_{10} (1500/1000) = 0.062$

   $\varepsilon_v = \Delta e/(1 + e_i) = 0.062/(1.900) = 0.0326$

2. **A** $\Delta H = H \varepsilon_v = 5(0.0326) = 0.163$ ft $= 1.96$ in

   Note that this is the same solution that would be obtained using the longer, combined equation above; it has merely been broken down to show Δe and ε_v.

Time Rate of Settlement

Applying a stress increase to a clay soil induces excess pore pressures, which dissipate with time in accordance with the theory of consolidation. For one-dimensional consolidation of a uniform clay layer, the average percent consolidation $U_{avg}(t)$ which has occurred by time t can be expressed as

$$U_{avg}(t) = f(T)$$

where

$$T = \frac{C_v t}{H_{dr}^2}$$

and

- T is a non-dimensional time factor from which $U_{avg}(t)$ can be determined.
- C_v is the coefficient of consolidation, with units of length^2/time.
- H_{dr} is the maximum length of drainage path in the clay layer. For a *doubly drained* layer (e.g., sand above and below the clay), it is one-half the clay layer thickness. For a *singly drained* layer (e.g., impervious shale under the clay), it is the same as the clay layer thickness. The subscript is used to emphasize that H_{dr} is not always equal to the layer thickness H (the NCEES Handbook uses the notation H in both cases).

The relationship $U_{avg} = f(t)$ is tabulated in Table 1.2.

Table 1.2 Relationship between U_{avg} and T

U_{avg}, %	T	U_{avg}, %	T
0 %	0.000	70	0.403
10	0.008	75	0.477
20	0.031	80	0.567
30	0.071	85	0.684
40	0.126	90	0.848
50	0.197	95	1.129
55	0.238	99	1.781
60	0.287	100	∞
65	0.340		

The settlement $S(t)$ at any time t is equal to the total calculated settlement multiplied by the degree of consolidation at time t.

Example 1.20

A 10 ft thick clay layer is drained by sand layers at the top and bottom. It is predicted to settle a total of 2.0 inches under the stress increase due to a foundation load. The coefficient of consolidation is 0.03 ft²/day.

1. The time required for 50% of the total settlement to occur is most nearly:
   (A) 640 days
   (B) 320 days
   (C) 160 days
   (D) 80 days

2. The time required for 90% of the total settlement to occur is most nearly:
   (A) 700 days
   (B) 2800 days
   (C) 1400 days
   (D) 350 days

3. Determine the settlement at the end of one year.
   (A) 0.73 in
   (B) 1.45 in
   (C) 2.00 in
   (D) 0.37 in

Solutions:

1. **C** As the 10 ft thick clay layer is bounded by a sand deposit on top and bottom, it is doubly drained and $H_{dr} = 10/2 = 5$ ft. For 50% consolidation, $T = 0.197$. Hence, $t_{50} = T_{50}H_{dr}^2/C_v = (0.197)(5^2)/0.03 = 164$ days.

2. **A** $t_{90} = T_{90}H_{dr}^2/C_v = (0.848)(5^2)/0.03 = 706$ days.

3. **B** $T = C_v t/H_{dr}^2 = (0.03)(365)/5^2 = 0.438$. For this value, $U_{avg} = 72.5$ %.
   $S(t) = 72.5\%(2 \text{ in}) = 1.45$ in.

1.8 Shear Strength of Soils

The Direct Shear Test

The shear strength of soil τ_f on any plane is a function of the normal stress on the plane. A simple test to measure shear strength is the *direct shear test*, illustrated on the left side of Figure 1.7. The soil is placed in a split box, and a vertical normal force V is applied. The horizontal force H needed to shear the soil is measured. If V and H are divided by the sample area A the normal stress σ and the shear stress τ_f are obtained. Plotting τ_f vs σ for samples tested at several normal stresses, as shown on the right of the figure, provides the strength parameters c and ϕ. The parameter c is commonly termed the *cohesion* and the parameter ϕ is termed the *angle of internal friction*, or simply the *friction angle*.

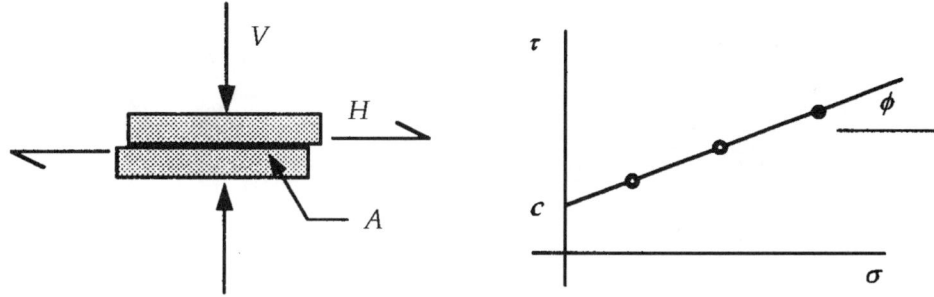

Figure 1.7 Direct Shear Test

The strength of soil along the failure plane is represented by the *Mohr-Coulomb strength equation*:

$$\tau_f = c + \sigma \tan \phi$$

where σ is the normal stress on the failure plane. Soil strength parameters may be defined in terms of *total stresses*, as written above, or in terms *of effective stresses*, as

$$\tau_f = c' + \sigma' \tan \phi'$$

The total stress representation is often used for short-term or undrained loading of saturated clays, for which $\phi = 0$, hence

$$\tau_f = c$$

For cohesionless sands, $c = 0$, and drained shear conditions (modeled with effective stress parameters) will usually prevail due to the relatively high permeability, hence:

$$\tau_f = \sigma' \tan \phi'$$

Nevertheless, one may often find the equation above written without the prime superscripts. For example, for dry sands, the total and effective stresses are the same.

Example 1.21

A sled with a rough bottom is loaded to a weight of 200 lb and pulled across a beach. It takes a horizontal force of 125 lb to pull the sled. What are the strength parameters of the beach sand?

(A) $c = 125$ lb/ft^2, $\phi = 0$
(B) $c = 0$, $\phi = 32°$
(C) $c = 125$ lb/ft^2, $\phi = 32°$
(D) $c = 200$ lb/ft^2, $\phi = 32°$

Solution:

As the material is sand, $c = 0$. The shear stress on the base of the sled is $\tau = 125/A$, where A is the area of the sled. The normal stress on the base of the sled is $\sigma = 200/A$. Applying the Mohr-Coulomb equation:

$$\tau_f = c + \sigma \tan \phi$$

$$\frac{125}{A} = 0 + \frac{200}{A} \tan \phi$$

$\therefore \tan \phi = 125/200 = 0.625$ and $\phi = 32°$

The answer is **B**.

Example 1.22

A sand for which $\phi = 35°$ is tested in direct shear in a 3 inch by 3 inch square shear box under a normal load of 200 lb.

1. Determine the normal stress on the sample.
   (A) 200 psi
   (B) 22.2 psi
   (C) 28.3 psi
   (D) 35 psi

2. Determine the shear stress on the sample at failure.
   (A) 15.6 psi
   (B) 22.2 psi
   (C) 70.0 psi
   (D) 28.3 psi

3. Determine the shear force on the sample at failure.
   (A) 22.2 lb
   (B) 70 lb
   (C) 200 lb
   (D) 140 lb

Solutions:

1. **B** $\sigma = P/A = 200/(3.0)^2 = 22.2$ psi
2. **A** $\tau_f = c + \sigma \tan \phi = 0 + 22.2 \tan 35° = 15.6$ psi
3. **D** $H = \tau A = (15.6)(9) = 140.4$ lb
   Alternatively, $H = V \tan \phi = 200 \tan 35° = 140.0$ lb (difference due to three-digit rounding).

The Unconfined Compression Test

The triaxial test permits more control over applied stress conditions than the direct shear test. A cylinder of soil is subjected to an all-around confining stress σ_3 and a deviator stress $(\sigma_1 - \sigma_3)$ applied vertically until failure (see Figure 1.8). The stress conditions on the sample at failure can be determined for all planes by plotting a Mohr's circle. If the test is repeated for soil samples at different confining stresses, the Mohr-Coulomb strength envelope can be obtained by drawing a line tangent to the failure circles as shown on the right side of the figure.

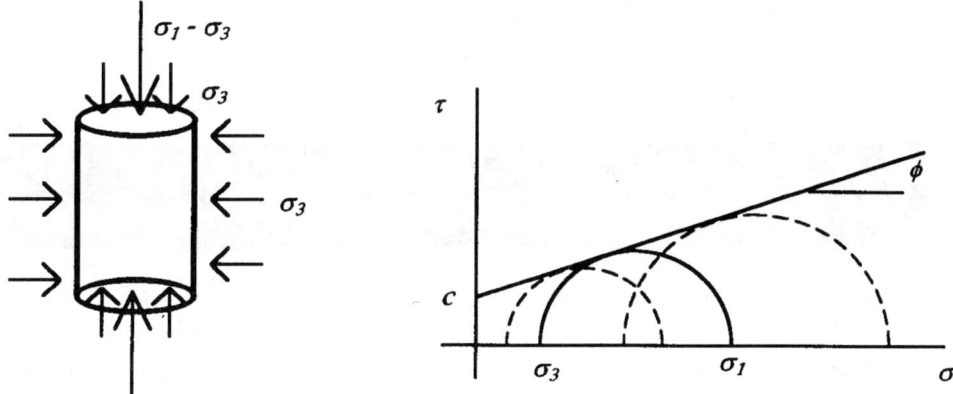

Figure 1.8 Triaxial Test, Mohr's Circles and Strength Envelope.

For saturated clays where $\phi = 0$, it is convenient to test without any confining stress ($\sigma_3 = 0$). In the resulting *unconfined compression test*, the normal stress σ_1 required for shear is termed the *unconfined compressive strength* q_u and is equal to twice the cohesion:

$$q_u = 2c$$

Example 1.23

A 2.0 inch diameter by 4.0 inch high cylindrical sample of clay is tested in an unconfined compression test. The sample fails under a force of 200 lb.

1. What is the normal stress on the sample at failure?
   (A) 200 psi
   (B) 50.0 psi
   (C) 63.7 psi
   (D) 15.9 psi

2. What is the unconfined compressive strength of the sample?
   (A) 200 psi
   (B) 50.0 psi
   (C) 63.7 psi
   (D) 15.9 psi

3. What is the cohesion of the sample?
   (A) 31.8 psi
   (B) 63.7 psi
   (C) 127.4 psi
   (D) 200 psi

Solutions:

1. **C** $\sigma = P/A = 200/(\pi\, 2.0^2/4) = 63.67$ psi
2. **C** The unconfined compression strength q_u is equal to the normal stress at failure, 63.7 psi.
3. **A** $c = q_u/2 = 63.67/2 = 31.8$ psi

1.9 Bearing Capacity of Soils

The basis of bearing capacity calculations is shown in Figure 1.9. When the applied pressure on the base of a footing reaches the *ultimate bearing capacity* q_{ult}, shear failure occurs along a curved surface below and to the sides of the footing. In the Terzaghi bearing capacity theory, the soil above the footing is replaced by a surcharge pressure γD. Soil below the footing has the properties c, ϕ, and γ.

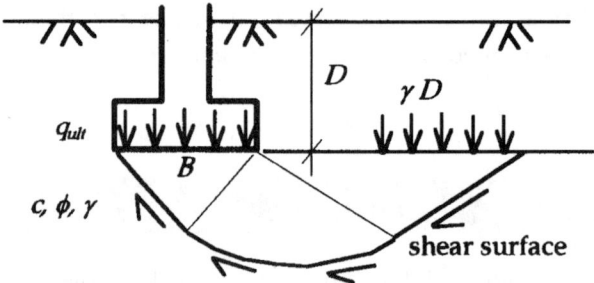

Figure 1.9 Bearing Capacity Failure under Footing

According to Terzaghi's theory, the ultimate bearing capacity q_{ult} for a concentrically loaded continuous footing of width B can be expressed as:

$$q_{ult} = cN_c + \gamma D N_q + 0.5\, B\gamma N_\gamma$$

where

c is the cohesion of the soil below the base of the footing.

γD is the vertical effective stress at the elevation of the footing base (sometimes also denoted q).

B is the footing width.

γ is the unit weight of the soil below the footing base.

N_c, N_q and N_γ are bearing capacity factors which are functions of the friction angle ϕ of the soil below the footing. These functions are widely tabulated and plotted, and there are several versions, reflecting different assumptions in their derivation.

Refinements and extensions of the above equation are available for more general conditions.

Example 1.24

A continuous footing is founded 3 ft below the ground surface in a clay for which $\gamma = 125$ lb/ft^3, $c = 1200$ lb/ft^2, and $\phi = 0$. For $\phi = 0$, assume $N_c = 5.14$, $N_q = 1$ and $N_\gamma = 0$.

1. Determine the ultimate bearing capacity.
   - (A) 1200 lb/ft^2
   - (B) 1575 lb/ft^2
   - (C) 6168 lb/ft^2
   - (D) 6540 lb/ft^2

2. If the factor of safety FS is to be at least 3, recommend an allowable bearing capacity q_a.
   - (A) 2180 lb/ft^2
   - (B) 19,350 lb/ft^2
   - (C) 6540 lb/ft^2
   - (D) 3600 lb/ft^2

Solutions:
1. **D** $q_{ult} = cN_c + \gamma D N_q + 0.5B\gamma N_\gamma = (1200)(5.14) + (125)(3\text{ ft})(1) + 0 = 6540$ lb/ft^2
2. **A** $q_a = q_{ult} / FS = 6450/3 = 2180$ lb/ft^2

Example 1.25

A continuous footing two feet wide is founded 3 ft below the ground surface in a uniform cohesionless sand. The soil parameters are $\gamma = 125$ lb/ft^3, $c = 0$, and $\phi = 34°$. This corresponds to $N_q = 29.4$ and $N_\gamma = 41.1$.

1. Determine the ultimate bearing capacity.
   - (A) 11,040 lb/ft^2
   - (B) 16,200 lb/ft^2
   - (C) 5,160 lb/ft^2
   - (D) 125 lb/ft^2
2. Determine the ultimate bearing force per lineal foot of foundation.
   - (A) 16,200 lb/ft
   - (B) 64,800 lb/ft
   - (C) 32,400 lb/ft
   - (D) 8,100 lb/ft

Solutions:

1. **B** $q_{ult} = cN_c + \gamma D N_q + 0.5B\gamma N_\gamma = 0 + (3)(125)(29.44) + (.5)(2)(125)(41.06) = $ 16,200 lb/ft^2
2. **C** $Q_{ult} = q_{ult} \times Area = q_{ult}(B)(1) = 16,200(2)(1) = 32,400$ lb/lineal ft

1.10 Earth Pressure and Retaining Walls

As soils possess shear strength, the horizontal stress in a soil can be greater or less than the vertical stress, and is dependent on the strain conditions. A small rotation of a retaining wall away from its backfill will reduce the horizontal earth pressure to its minimum value, the *active earth pressure* condition. Retaining walls are usually designed for active earth pressure conditions. For the special but common case of a frictionless vertical wall backfilled with a cohesionless material with a horizontal backfill surface, the *active earth pressure coefficient* K_a is

$$K_a = \tan^2(45 - \phi/2)$$

For dry sand with no water table, the active earth pressure p_a at any depth z is equal to

$$p_a = K_a \sigma'_v = K_a \gamma z$$

Knowing the active earth pressure at any point along a face of a wall, the total earth force on a wall can be obtained by integrating the pressure with respect to depth, or calculating the area of an earth pressure vs. depth diagram.

Example 1.26

The retaining wall shown is 8 ft high and retains a clean dry sand for which $\gamma = 120$ lb/ft^3 and $\phi = 32°$.

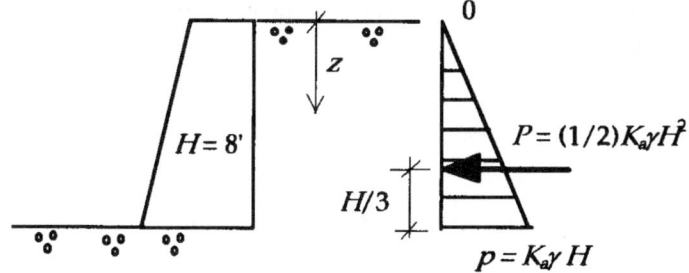

1. Calculate the active earth pressure at the base of the wall.
   (A) 960 lb/ft^2
   (B) 480 lb/ft^2
   (C) 295 lb/ft^2
   (D) 1180 lb/ft^2

2. Calculate the total earth force on the wall.
   (A) 960 lb/ft
   (B) 480 lb/ft
   (C) 295 lb/ft
   (D) 1180 lb/ft

3. Calculate the overturning moment per lineal foot of wall due to the active earth pressure.
   (A) 1180 ft-lb
   (B) 9440 ft-lb
   (C) 1280 ft-lb
   (D) 3140 ft-lb

Solutions:

1. **C** $K_a = \tan^2(45 - \phi/2) = \tan^2(45 - 32/2) = 0.307$
   $p_a = K_a \gamma z = (0.307)(120)(8) = 295$ lb/ft^2

2. **D** The total active earth force is the integral of the earth pressure from $z = 0$ to 8':
   $P_a = 0.5 K_a \gamma H^2 = (0.5)(0.307)(120)(8)^2 = 1180$ lb/ft of wall

3. **D** The active force acts at the centroid of the earth pressure diagram; in this case, $H/3$. Thus $M_o = P_a \times H/3 = (1179)(8/3) = 3140$ ft-lb/ft of wall

Note: If a force is applied on the outside of the wall, pushing it toward the backfill, the horizontal pressure increases until it reaches the *passive earth pressure condition*, at which shear failure occurs and the pressure and force can no longer increase. The *passive earth pressure coefficient* is

$$K_p = \tan^2(45 + \phi/2)$$

Example 1.27

For the same wall shown in the previous example, assume a force is applied to the outside until passive earth pressure conditions occur.

1. Calculate the passive earth pressure at the base of the wall.
   - (A) 3120 lb/ft²
   - (B) 295 lb/ft²
   - (C) 1560 lb/ft²
   - (D) 6240 lb/ft²

2. Calculate the total passive earth force on the wall.
   - (A) 6240 lb/ft
   - (B) 3120 lb/ft
   - (C) 12,480 lb/ft
   - (D) 24,960 lb/ft

Solutions:

1. **A** $K_p = \tan^2(45 + 32/2) = 3.25$
   $p_a = K_p \gamma z = (3.25)(120)(8) = 3120 \text{ lb/ft}^2$

2. **C** The total active earth force is the integral of the earth pressure from $z = 0$ to 8':
   $P_p = 0.5 K_p \gamma H^2 = (0.5)(3.25)(120)(8)^2 = 12,480 \text{ lb/ft of wall}$

1.11 Slope Stability

A simple approach to assessing the stability of a slope is to consider the statics on a wedge of soil such as ABC shown in Figure 1.10.

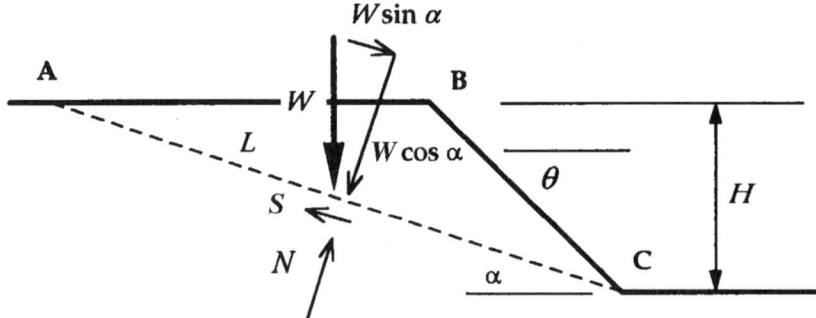

Figure 1.10 Free Body Considered for Slope Stability Analysis

If the wedge slides along a plane AC inclined at angle α, the applied shear force parallel to the plane is $W \sin \alpha$. The normal force on plane AC is $N = W \cos \alpha$. If the soil has strength parameters c and ϕ, the maximum shear force S that can be developed along plane AC is

$$S = cL + N \tan \phi = cL + W \cos\alpha \tan \phi$$

If the factor of safety against sliding is taken as the ratio of the available shear force to the applied shear force, one obtains the equation from the NCEES Handbook:

$$FS = (cL + W \cos\alpha \tan \phi)/W \sin\alpha$$

The critical angle α must be found by trial and error, and would be the value yielding the lowest factor of safety. In practice, non-planar surfaces, especially circular surfaces, are also considered, using more rigorous methods.

Example 1.28

Refer to Figure 1.9. Assume $H = 20$ ft, $\gamma = 120$ lb/ft^3, $c = 500$ lb/ft^2, and $\phi = 30°$. Assume the slope angle $\theta = 26.6°$ (a 1 on 2 slope) and the failure surface angle $\alpha = 15°$. Determine the factor of safety FS.

(A) 3.2
(B) 5.7
(C) 1.00
(D) 1.76

Solution:

B The length AB = $(H/\tan\alpha) - (H/\tan\theta) = (20/0.2679) - (20/0.5) = 34.65$ ft

The weight of the soil wedge is

$$W = Area \times \gamma$$
$$= (0.5)(20)(34.65)(120) = 41{,}580 \text{ lb/ft}^2$$

The length $L = 20/\sin\alpha = 77.27$ ft. Then

$$FS = (cL + W\cos\alpha \tan\phi)/W\sin\alpha$$
$$= [500 \times 77.27 + 41{,}580 \cos 15° \tan 30°]/41{,}580 \sin 15° = 5.7$$

2. Structural Analysis and Design

by Frank J. Hatfield

The material regarding structural analysis and design given on pages 87, 89-92, and 97 of the Handbook, 3rd ed., is an arbitrary sampling of particular methods and formulas. The following review is based on the hope that the examination itself will focus on the same specific topics as well as on basic concepts.

2.1 Analysis of Determinate Trusses

A truss is a structure in which all members are straight, all joints are hinges (i.e., cannot transmit a moment), and all loads are applied at joints. Therefore, there are no moments in the members, and member forces are directed axially. The virtual work formula for truss deflection is on page 87 of the Handbook. However, in the definitions of terms following the formula, the definitions of F_Q and F_p are erroneously interchanged.

Example 2.1

The figure shown represents a plane truss with a load.

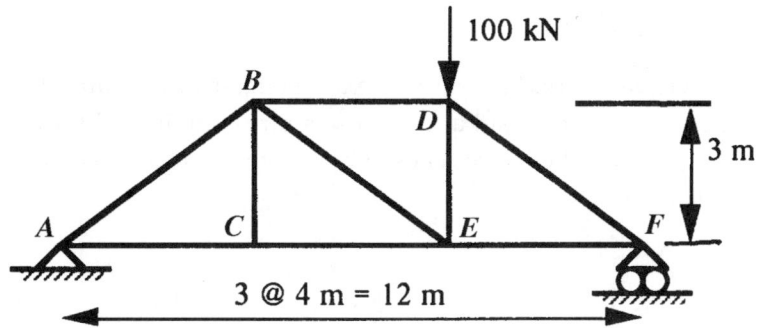

1. What is the reaction force at support F?
   (A) −33 kN
   (B) 0 kN
   (C) 33 kN
   (D) 67 kN

2. What is the force in member BC?
   (A) 0 kN
   (B) 100 kN tension
   (C) 100 kN compression
   (D) 133 kN tension

3. What is the force in member CE?
   (A) 0 kN
   (B) 44 kN tension
   (C) 44 kN compression
   (D) 89 kN tension

4. Assume that for all members, cross-sectional area is 900 mm² and modulus of elasticity is 200 GPa. What is the horizontal deflection at joint F?
   (A) 0 mm
   (B) 1 mm
   (C) 2 mm
   (D) 4 mm

Solutions:

1. **D** Sum of moments about A is zero:

   $12 R_F - 8 \text{ m} \times 100 \text{ kN} = 0$. $R_F = 67$ kN.

2. **A** Cut members AC, BC, CE and investigate the free-body containing joint C. Sum of vertical forces is zero. $F_{BC} = 0$ kN.

3. **B** Cut members BD, BE, CE and investigate the free-body containing joints D, E, F. Sum of moments about B is zero:

   $8 R_F - 4 \text{ m} \times 100 \text{ kN} - 3 F_{CE} = 0$.

   Since $R_F = 67$ kN, then

   $F_{CE} = 44$ kN.

4. **D** Use virtual work. First compute all member forces caused by real load. Then remove real load and impose a virtual unit load horizontally at F. Compute member forces caused by the virtual load. $F_{QAC} = F_{QCE} = F_{QEF} = 1$; all other $F_Q = 0$:

   $\Delta = \Sigma F_Q F_p L / AE$

   $= (1 \times 44 \text{ kN} \times 4 \text{ m} + 1 \times 44 \times 4 + 1 \times 89 \times 4)$
   $\times 1000 / (900 \text{ mm}^2 \times 200 \text{ kN/mm}^2)$

   $= 4$ mm

2.2 Analysis of Determinate Beams

A beam is a member subjected to lateral forces which cause internal shear and moment. The Handbook, on page 97, gives a table of fixed-end moments which suggests that the examination may include an indeterminate beam, and that it will be most conveniently analyzed by a pre-computer method called moment distribution. However, a determinate beam was chosen for the example that follows because in review it is most important to focus on basic concepts. The virtual work formula for beam deflection given on page 87 of the Handbook will be unfamiliar to many readers. It may be better to review a familiar method rather than learning the virtual work method. A definition of influence lines is given on page 87 of the Handbook.

Example 2.2

The figure shown represents a beam with an internal hinge at B and a moving load.

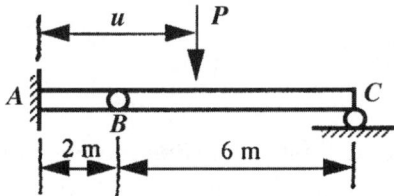

1. What is the ordinate of the influence line for the reaction force at C for $u = 1.0$ m?
   (A) 0
   (B) 0.25
   (C) 0.50
   (D) 1.0

2. What is the shear at B for $P = 60$ kN and $u = 5$ m ?
   (A) 0 kN
   (B) 30 kN
   (C) 40 kN
   (D) 60 kN

3. What is the maximum moment on segment BC for $P = 60$ kN ?
   (A) 0 kN-m
   (B) 10 kN-m
   (C) 60 kN-m
   (D) 90 kN-m

4. Assume the moment of inertia is 300×10^6 mm^4 and modulus of elasticity is 200 GPa. What is the angular rotation in radians at B on segment AB for $P = 60$ kN and $u = 2$ m ?
   (A) 0.001
   (B) 0.002
   (C) 0.004
   (D) 0.010

Solutions:

1. **A** Influence line is a graph of the value of R_C as a function of load position u, with $P = 1$. When $u \leq 2$, $R_C = 0$.

2. **B** Use segment BC as free-body. Sum of moments about C is zero:

   $$3 \text{ m} \times 60 \text{ kN} - 6 V_B = 0. \therefore V_B = 30 \text{ kN}.$$

3. **D** Locate P at $u = 5$ m to maximize moment. Use right half of segment BC as free-body. Sum of moments about point at middle of BC is zero:

   $$3 R_C - M = 0.$$

   Since $R_C = 30$ kN, then

   $$M = 90 \text{ kN-m}.$$

4. **B** Use virtual work. First determine moment function for real load. $M = 60x - 120{,}000$ kN-mm for segment AB; $M = 0$ for segment BC. Then remove real load and impose a virtual unit moment at B. Determine the moment function for the virtual load; $m = 1$ for segment AB; $m = 0$ for segment BC:

   $$\Delta = \int (mM/EI)\, dx$$

   $$= \int 1\,(60x - 120{,}000)\, dx\, /\, (300 \times 10^6 \times 200)$$

   $$= -120 \times 10^6\, /\, (300 \times 10^6 \times 200) = 0.002 \text{ radians}$$

   Note: Δ can represent any type of deformation; here it represents rotation.

2.3 Design in Reinforced Concrete

The material on Concrete Design of the Handbook is selected from a specification published by the American Concrete Institute. A review of slabs, beams and columns is suggested.

Example 2.3

The figure shown represents the profile and cross-section of a reinforced concrete beam. Reinforcing is Grade 60 steel, and the 28-day compressive strength of the concrete is 3000 psi.

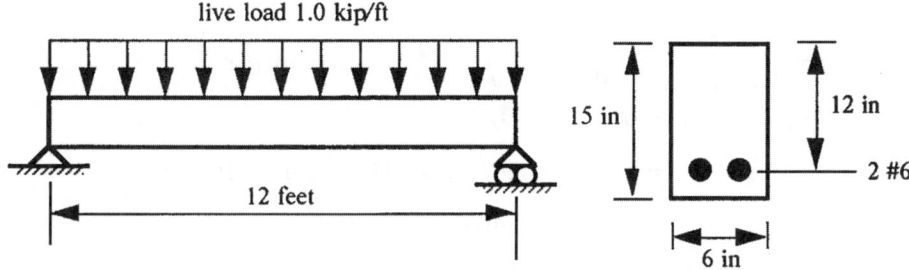

1. The beam is...
   (A) Over-reinforced
   (B) Under-reinforced
   (C) Balanced
   (D) Doubly reinforced

2. What is the design moment strength?
   (A) 20 kip-ft
   (B) 33 kip-ft
   (C) 41 kip-ft
   (D) 53 kip-ft

3. What is the required moment strength?
   (A) 15 kip-ft
   (B) 20 kip-ft
   (C) 24 kip-ft
   (D) 33 kip-ft

4. Shear reinforcement is...
   (A) not needed at the mid-span of the beam
   (B) not needed at the ends of the beam
   (C) not needed anywhere on the beam
   (D) needed for the full length of the beam

Solutions:

1. **B** $\rho = A_s/bd = 2 \times 0.44 / 6 \times 12 = 0.012$

    $$\rho_b = \frac{0.85\,\beta f_c'}{f_y} \cdot \frac{87000}{87000 + f_y}$$

    $$= \frac{0.85 \times 0.85 \times 3000}{60000} \times \frac{87000}{87000 + 60000} = 0.021$$

    $\rho < \rho_b$ so beam is under-reinforced

2. **C** $a = \dfrac{A_s f_y}{0.85 f_c' b} = \dfrac{2 \; 0.44 \times 60000}{0.85 \times 3000 \times 6} = 3.45$ inches

    $\phi M_N = \phi A_s f_y (d - a/2)$

    $= 0.90 \times 2 \times 0.44 \times 60000 \, (12 - 3.45/2)$

    $= 488 \times 10^3$ lb-in or 41 kip-ft

3. **D** $w_{dead} = (6/12)$ ft $\times$ $(15/12)$ ft $\times$ 150 pcf = 94 lb/ft

    $M_{dead} = w_{dead} L^2 / 8 = 0.094$ kip/ft $\times$ $(12$ ft$)^2 / 8 = 1.7$ kip-ft

    $M_{live} = w_{live} L^2 / 8 = 1.0$ kip/ft $\times$ $(12$ ft$)^2 / 8 = 18$ kip-ft

    $M_U = 1.4\, M_{dead} + 1.7\, M_{live} = 1.4 \times 1.7 + 1.7 \times 18 = 33$ kip-ft

4. **A** Shear requirement computed at distance d from support.

    $V_{dead} = 5$ ft $\times$ 0.094 kip/ft = 0.5 kip

    $V_{live} = 5$ ft $\times$ 1.0 kip/ft = 5.0 kip

    $V_U = 1.4\, V_{dead} + 1.7\, V_{live} = 1.4 \times 0.5 + 1.7 \times 5.0 = 9.2$ kip

    $V_C = 2\sqrt{f_c'}\, b\, d = 2\sqrt{3000} \times 6 \times 12 = 7890$ lb or 7.9 kips

    $\phi V_C / 2 = 0.85 \times 7.9 / 2 = 3.4$ kips

    Since $V_U > \phi V_C / 2$ shear reinforcement is needed near the supports. However V_U decreases with distance from the support, and is less than $\phi V_C / 2$ near midspan. Therefore, there is a section of the beam near the middle where shear reinforcement is not needed.

2.4 Design in Structural Steel

The material on Steel Design of the Handbook is selected from specifications published by the American Institute of Steel Construction. Both the ASD (allowable stress design) and the LRFD (load and resistance factor design) specifications are included. Since LRFD is the more modern methodology and is the one taught at nearly all colleges and universities, it is the only one that will be covered in the following example. The Handbook includes obsolete evaluations of the LRFD shear lag factor U for tension members and of the LRFD block shear design strength for bolted connections. These formulas are from the first edition LRFD specification and differ from those in the current second edition LRFD specification. For the purposes of passing the FE examination, it probably is safest to use the obsolete formulas given in the Handbook.

Example 2.4

The sketch shown represents the end of a double-angle steel tension member with holes for a bolted connection. Yield strength of the steel is 50 ksi, ultimate tensile strength is 65 ksi. The gross cross-sectional area of the two angles together is 2.88 in². Holes are punched 13/16 diameter for 3/4 inch bolts.

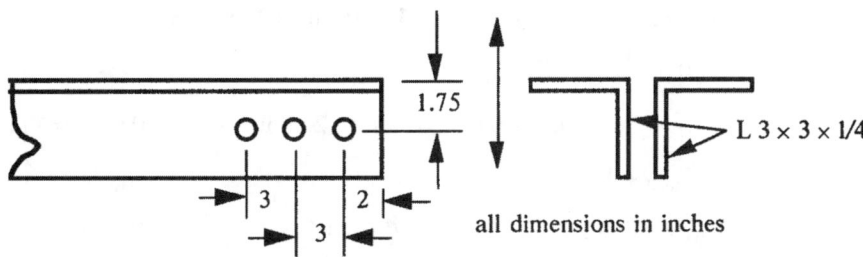

all dimensions in inches

1. What is the design strength for yield?
    (A) 101 kips
    (B) 111 kips
    (C) 130 kips
    (D) 264 kips

2. What is the design strength for tension fracture?
    (A) 101 kips
    (B) 111 kips
    (C) 130 kips
    (D) 264 kips

3. What is the design strength for block shear fracture?
   (A) 101 kips
   (B) 111 kips
   (C) 130 kips
   (D) 264 kips

4. Assume that the critical design strength is 96 kips and that dead load is 40 kips. What is the maximum permissible live load?
   (A) 30 kips
   (B) 35 kips
   (C) 40 kips
   (D) 48 kips

Solutions: (LRFD)

1. **C** $\phi F_y A_g = 0.9 \times 50 \text{ ksi} \times 2.88 \text{ in}^2 = 130 \text{ kips}$

2. **A** $A_n = 2.88 - 2 \times 1/4 \text{ in} \times 7/8 \text{ in} = 2.44 \text{ in}^2$

    $A_e = U A_n = 0.85 \times 2.44 \text{ in}^2 = 2.07 \text{ in}^2$ (first edition LRFD)

    $\phi F_u A_e = 0.75 \times 65 \text{ ksi} \times 2.07 \text{ in}^2 = 101 \text{ kips}$

3. **B**   Using first edition LRFD

    $A_{gt} = 2 \times 1/4 \text{ in} \times (3 - 1.75) \text{ in} = 0.625 \text{ in}^2$

    $A_{nv} = 2 \times 1/4 \times (8 - 2.5 \times 13/16) \text{ in} = 2.98 \text{ in}^2$

    $\phi (0.6 F_u A_{nv} + F_y A_{gt}) =$

    $0.75 (0.6 \times 65 \text{ ksi} \times 2.98 \text{ in}^2 + 50 \times 0.625) = 111 \text{ kips}$

4. **A** $T_u = 1.2 T_{dead} + 1.6 T_{live}$

    $T_{live} = (T_u - 1.2 T_{dead})/1.6 = (96 - 1.2 \times 40)/1.6 = 30 \text{ kips}$

3. Transportation

by Thomas Maleck

Many of the questions relating to transportation can be answered by using the material found in the American Association of State Highway and Transportation Officials (AASHTO) book "A Policy on Geometric Design of Highways and Streets, 1984". This guide is often referred to as the "GREEN BOOK". Additional help can be found in *Principles & Practice of Civil Engineering* by Merle C. Potter.

3.1 Braking Distance (stopping sight distance)

Please refer to chapter three of the AASHTO's "A Policy on Geometric Design of Highways and Streets, 1984" to obtain the details needed to properly understand this material. See page 87 of the NCEES Reference Handbook, 3rd ed., for the equation used to calculate braking distance.

Example 3.1

A two lane highway has a grade of 2 percent. The design speed is 60 mph. Refer to the following table.

Design Speed (mph)	Assumed Speed for Condition (mph)	Brake Reaction Time (sec)	Brake Reaction Distance (ft)	Coefficient of Friction f	Braking Distance on Level[a] (ft)	Stopping Sight Distance Computed[a] (ft)	Stopping Sight Distance Rounded for Design (ft)
50	44-50	2.5	161.3-183.3	0.30	215.1-277.8	376.4-461.1	400-475
55	48-55	2.5	176.0-201.7	0.30	256.0-336.1	432.0-537.8	450-550
60	52-60	2.5	190.7-220.0	0.29	310.8-413.8	501.5-633.8	525-650
65	55-65	2.5	201.7-238.3	0.29	347.7-485.6	549.4-724.0	550-725
70	58-70	2.5	212.7-256.7	0.28	400.5-583.3	613.1-840.0	625-850

[a]Different values for the same speed result from using unequal coefficients of friction.

1. What is the assumed driver reaction time?
   (A) 1.0 sec
   (B) 1.5 sec
   (C) 2.0 sec
   (D) 2.5 sec

2. What is the lowest allowable assumed speed used for determining stopping sight distance?
   (A) 52 mph
   (B) 55 mph
   (C) 57 mph
   (D) 60 mph

3. What is the assumed coefficient of friction?
   (A) 0.28
   (B) 0.29
   (C) 0.30
   (D) 0.31

4. What is the expected braking distance?
   (A) 405 ft
   (B) 425 ft
   (C) 445 ft
   (D) 465 ft

5. How far does the vehicle travel before the vehicle decelerates?
   (A) 220 ft
   (B) 230 ft
   (C) 240 ft
   (D) 250 ft

6. What is the required stopping sight distance?
   (A) 625 ft
   (B) 665 ft
   (C) 695 ft
   (D) 715 ft

Solutions:

1. **D** From the table, brake reaction time equals driver reaction time, which equals 2.5 sec.

2. **A** From the table, assume speed for condition is 52-60 mph. For wet pavement 52 mph is assumed for design.

3. **B** From the table, coefficient of friction for a design speed of 60 mph is 0.29.

4. **C** $d = \dfrac{V^2}{2g(f \pm s)}$

   $V = 60$ mph, which $= 88$ fps, $g = 32.2$, $f = 0.29$, $s = -0.02$. s is negative because one direction of travel is going downhill.

   Therefore, $d = 445.36$ ft.

5. **A** Distance vehicle travels before braking is $T \times V$ where

   $T = 2.5$ sec (driver reaction)

   $V = 60$ mph $= 88$ fps

   Therefore, distance $= 2.5(88) = 220$ ft.

6. **B** Stopping sight distance = braking distance + distance the vehicle travels before braking, which equals questions No. 4 plus No. 5:

   $$445 + 220 = 665 \text{ ft}$$

3.2 Sight Distance Over a Vertical Curve

The equations needed to solve the problem of sight distance over a vertical curve are found on page 87 of the NCEES Reference Handbook.

Example 3.2

A freeway has a 70 mph design speed. There is a 2% grade followed by a negative 3% grade. (Hint: this is a crest vertical curve.) Assume height of driver's eye to be 3.5 ft and object height to be 0.5 ft. Refer to the table given in Example 3.1, as well as the following table.

Design Speed (mph)	Time(s) Premaneuver Detection & Recognition	Decision & Response Initiation	Maneuver (Lane Change)	Decision Sight Distance (ft) Summation	Computed	Rounded for Design
30	1.5-3.0	4.2-6.5	4.5	10.2-14.0	449- 616	450- 625
40	1.5-3.0	4.2-6.5	4.5	10.2-14.0	598- 821	600- 825
50	1.5-3.0	4.2-6.5	4.5	10.2-14.0	748-1,027	750-1,025
60	2.0-3.0	4.7-7.0	4.5	11.2-14.5	986-1,276	1,000-1,275
70	2.0-3.0	4.7-7.0	4.0	10.7-14.0	1,098-1,437	1,100-1,450

1. What is the required length of vertical curve needed to satisfy design stopping sight distance?
   (A) 1435 ft
   (B) 2720 ft
   (C) 4550 ft
   (D) 1670 ft

2. What is the required length of vertical curve needed to satisfy the minimum value for decision sight distance?
   (A) 1435 ft
   (B) 2720 ft
   (C) 4550 ft
   (D) 1670 ft

Solutions:

1. **B** From the table, stopping sight distance = 850 ft.

   For a crest vertical curve:

   $$S < L \quad L = \frac{AS^2}{100\left(\sqrt{2H_1} + \sqrt{2H_2}\right)^2}$$

   $$S > L \quad L = 2S - \frac{200\left(\sqrt{H_1} + \sqrt{H_2}\right)^2}{A}$$

   L = length of vertical curve
   A = algebraic difference in grade which equals $2 - (-3) = 5$
   S = sight distance requirement which equals 850 ft
   H_1 = height of driver's eye which equals 3.5 ft
   H_2 = height of object which equals 0.5 ft

   For $S > L$

   $$L = (2)(850) - \frac{(200)\left(\sqrt{3.5} + \sqrt{0.5}\right)^2}{5}$$

   $$= 1434.17 \text{ ft}$$

   False answer: S is <u>not</u> larger than L.

   For $S < L$

   $$L = \frac{(5)(850)^2}{(100)\left(\sqrt{7.0} + \sqrt{1}\right)^2}$$

   $$= 2717.90 \text{ ft}$$

   Correct answer: S is smaller than L.

2. **C** From the table, minimum value is 1,100 ft.

   A, H_1 and H_2 do not change.

   For $S > L$

   $$L = (2)(1100) - \frac{(200)\left(\sqrt{3.5} + \sqrt{0.5}\right)^2}{5}$$

   $$= 1934.17 \text{ ft}$$

   False answer: S is <u>not</u> larger than L.

   For $S < L$

   $$L = \frac{(5)(1100)^2}{(100)\left(\sqrt{7} + \sqrt{1}\right)^2}$$

   $$= 4551.78 \text{ ft}$$

   Correct answer: S is smaller than L.

3.3 Sight Distance for a Sag Vertical Curve

The sag vertical curve equations are also found on page 87 of the NCEES Handbook.

Example 3.3

A highway has a 55 mph design speed. There is a negative 1 percent grade followed by a 2 percent grade. Refer to the following table.

Design Speed (mph)	Assumed Speeds		Minimum Passing Sight Distance (ft)
	Passed Vehicle (mph)	Passing Vehicle (mph)	
50	41	51	1,840
60	47	57	2,140
65	50	60	2,310
70	54	64	2,490

1. What is the required length of vertical curve needed to satisfy AASHTO stopping sight distance?
   (A) 320 ft
   (B) 585 ft
   (C) 200 ft
   (D) 1460 ft

2. What is the required length of vertical curve needed to satisfy AASHTO passing sight distance?
   (A) 320 ft
   (B) 585 ft
   (C) 200 ft
   (D) 1460 ft

3. What is the required length of vertical curve needed to satisfy AASHTO design sight distance?
   (A) 320 ft
   (B) 585 ft
   (C) 200 ft
   (D) 1460 ft

4. What is the required length of vertical curve to satisfy AASHTO requirements for comfort?
   (A) 320 ft
   (B) 585 ft
   (C) 200 ft
   (D) 1460 ft

Solutions:

1. **A** From the table in Example 3.1, stopping sight distance = 550 ft.
$$A = (-1) - 2 = -3 \text{ or } 3$$
For a sag vertical curve:

$$S > L \qquad L = 2S - \frac{400 + 3.55S}{A}$$

$$S < L \qquad L = \frac{AS^2}{400 + 3.55S}$$

For $S > L$
$$L = 2(550) - \frac{400 + 3.55(550)}{3}$$
$$= 315.83 \text{ ft}$$

Correct Answer: S is greater than L.

For $S < L$
$$L = \frac{3(550)^2}{400 + 3.55(550)}$$
$$= 385.76 \text{ ft}$$

False answer: S is <u>not</u> less than L.

2. **D** From the table given above, passing sight distance = 1950 ft for 55 mph. A still equals 3.

For $S > L$
$$L = 2(1950) - \frac{400 + 3.55(1950)}{3}$$
$$= 1459.17 \text{ ft}$$

Correct answer: S is greater than L.

For $S < L$
$$L = \frac{3(1950)^2}{400 + 3.55(1950)}$$
$$= 1557.9 \text{ ft}$$

False answer: S is <u>not</u> smaller than L.

3. **B** From the table of Example 3.2, decision sight for 55 mph is 875 to 1150 ft. Using a minimum value of 875 for this example. A still equals 3.

For $S > L$
$$L = 2(875) - \frac{400 + 3.55(875)}{3}$$
$$= 581.25 \text{ ft}$$

Correct answer: S is greater than L.

For $S < L$
$$L = \frac{3(875)^2}{400 + 3.55(1950)}$$
$$= 313.67 \text{ ft}$$

False answer: S is <u>not</u> less than L.

4. **C** $L = \dfrac{AV^2}{46.5}$

   $A = 3$ and $V = 55$ mph.

   Therefore,

   $$L = \dfrac{3(55)^2}{46.5}$$
   $$= 195.16 \text{ ft}$$

3.4 Vertical Curve Elevations

Refer to page 96 of the NCEES Reference Handbook for a figure and equations used for calculations involving vertical curves.

Example 3.4

An 800 ft vertical curve with equal legs is provided for a highway crest vertical curve. The grade of the back tangent is 2 % and the grade of the forward tangent is a negative 2 %. The elevation of the Point of Vertical Intersection (*PVI*) is 1200 ft.

1. What is the elevation of the Point of Vertical Curve (*PVC*)?
   (A) 1192 ft
   (B) 1208 ft
   (C) 1184 ft
   (D) 1216 ft

2. What is the elevation of the Point of Vertical Tangent (*PVT*)?
   (A) 1192 ft
   (B) 1208 ft
   (C) 1184 ft
   (D) 1216 ft

3. What is the Rate of Change of Grade?
   (A) −0.050
   (B) −0.500
   (C) 0.01
   (D) 0.10

4. What is the Tangent Offset at the *PVI*?
   (A) −0.40
   (B) −2.00
   (C) −4.00
   (D) 0.20

5. If the *PVI* is at station 11+00 and the *PVC* is at station 7+00, what is the Tangent Offset at station 8+00?
   (A) 0.25
   (B) 0.50
   (C) 1.00
   (D) −0.25

6. What is the Curve Elevation at station 8+00?
   (A) 1193.75 ft
   (B) 1194.00 ft
   (C) 1194.50 ft
   (D) 1195.00 ft

Solutions:

1. **A** $PVC = PVI - (g_1)L/2$
   $$= 1200 - (2)8/2 = 1192$$
   (*L* is in stations)

2. **A** $PVT = PVI + (g_2)L/2$
   $$= 1200 + (-2)8/2 = 1192$$

3. **B** Rate of Change:
   $$r = \frac{g_2 - g_1}{L} = \frac{(-2) - 2}{8}$$
   $$= -0.500$$

4. **C** a = Parabola Constant
   $$a = \frac{g_2 - g_1}{2L} = \frac{(-2) - 2}{2(8)}$$
   $$= -0.25$$

   E = Tangent Offset at *PVI*
   $$E = a\left(\frac{L}{2}\right)^2 = (-0.25)\frac{8^2}{2^2}$$
   $$= -4.00$$

5. **D** y = Tangent Offset
   $$y = ax^2 = (-0.25)(1)^2$$
   $$= -0.25$$
   (*x* is in stations)

6. **A** Curve Elevation = $Y_{PVC} + g_1 x + ax^2$
   $Y_{PVC} = 1192$ ft
   $g_1 = 2\%$

$x = 1$ station
$a = -0.25$

Therefore: Curve Elevation at station 8+00 equals:

$$1192 + 2(1) - 0.25(1)^2 = 1193.75$$

3.5 Horizontal Curves

The horizontal curve figure and equations are found on page 97 of the NCEES Handbook.

Example 3.5

A horizontal curve with a 1000 ft radius is to be provided for two perpendicular roads. The station of the *PI* is 12+00.00.

1. What is the Intersection of Angle?
    (A) 90°
    (B) 120°
    (C) 45°
    (D) 60°

2. What is the Degree of Curve?
    (A) 5.73°
    (B) 5.00°
    (C) 4.25°
    (D) 6.81°

3. What is the Tangent Distance?
    (A) 800 ft
    (B) 550 ft
    (C) 1000 ft
    (D) 1100 ft

4. What is the External Distance?
    (A) 411.8 ft
    (B) 414.2 ft
    (C) 618.2 ft
    (D) 411.2 ft

5. What is the Length of the Middle Ordinate?
    (A) 292 ft
    (B) 293 ft
    (C) 288 ft
    (D) 283 ft

72   Ch. 3 / Transportation—Horizontal Curves

6. What is the Length of Curve?
   (A) 1200.00 ft
   (B) 1242.81 ft
   (C) 1500.00 ft
   (D) 1570.80 ft

7. What is the stationing of the PC?
   (A) 17+70.80
   (B) 15+81.00
   (C) 2+00.00
   (D) 2+70.80

8. What is the stationing of the PT?
   (A) 17+70.80
   (B) 15+81.00
   (C) 2+00.00
   (D) 2+70.80

Solutions:

1. **A** Since the roads are perpendicular, the Angle of Intersection is 90°.

2. **A** The Degree of Curve $= \dfrac{5729.58}{R}$

   $R = 1000$ ft. Therefore,
   $$D = \dfrac{5729.58}{1000} = 5.73°$$

3. **C** The Tangent Distance is
   $$T = R\tan(I/2)$$
   $I = 90°$ (from question 1). Therefore,
   $$T = (1000)\tan(90/2) = 1000.00 \text{ ft}$$

4. **B** The External Distance is
   $$E = R\left[(\sec I/2) - 1\right]$$
   $$= 1000[1.41421 - 1] = 414.21 \text{ ft}$$

5. **B** The length of the Middle Ordinate is
   $$M = R - R\cos\dfrac{I}{2}$$
   $R = 1000.00$
   $I = 90°$
   Therefore: $M = 292.89$ ft

6. **D** The Length of Curve is

$$L = \frac{100I}{D}$$

$$= \frac{100(90)}{5.72958} = 1570.80 \text{ ft}$$

7. **C** Station of PC = station of PI − Tangent Distance:

$$PC = 12+00.00 - 1000.00 = 2+00.00$$

8. **A** Station of PT equals station of PC plus Length of Curve:

$$PT = PC + L$$
$$= 2+00.00 + 1570.80$$
$$= 17+70.80$$

3.6 Superelevations

To calculate the superelevation of a highway, refer to the material on page 88 of the NCEES Handbook, 3rd ed. The superelevation equation for railroads is also presented there.

Example 3.6

For the horizontal curve in Example 3.5, what is the minimum superelevation needed to provide for a 50 mph design speed? Refer to the following table.

Design Speed (mph)	Maximum e	Maximum f	Total (e + f)	Maximum Degree of Curve	Rounded Maximum Degree of Curve	Minimum Radius (ft)
20	.08	.17	.25	53.54	53.5	107
30	.08	.16	.24	22.84	22.75	252
40	.08	.15	.23	12.31	12.25	468
50	.08	.14	.22	7.54	7.5	764
60	.08	.12	.20	4.76	4.75	1,206
65	.08	.11	.19	3.85	3.75	1,528
70	.08	.10	.18	3.15	3.0	1,910

NOTE: In recognition of safety considerations, use of e_{max} = 0.04 should be limited to urban conditions.

(A) 0.165

(B) 0.04

(C) 0.140

(D) 0.027

Solution:

The maximum allowable lateral friction force is required to minimize the needed superelevation. Therefore, from the given table, the maximum allowable friction factor f is 0.140.

The required superelevation e is equal to

$$e + f = \frac{V^2}{gR}$$

Therefore:

$$e = \frac{V^2}{gR} - f$$

$V = 50 \text{ mph} = 73.33 \text{ ft/sec}$

$R = 1000 \text{ ft}$

$g = 32.2 \text{ ft/sec}^2$

$$e = \frac{73.33^2}{32.2(1000)} - 0.14 = 0.027$$

The answer is **D**.

3.7 Spirals

Spirals for highways are analyzed using the equation given on page 88 of the NCEES Reference Handbook. Railroad spirals are also analyzed using the equation given there.

Example 3.7

For the horizontal curve in Examples 3.5 and 3.6, what is the recommended length of spiral transition?
 (A) 150 ft
 (B) 200 ft
 (C) 250 ft
 (D) 300 ft

Solution:

The length of spiral is equal to:

$$L_S = 1.6 \frac{V^3}{R}$$

$V = 50 \text{ mph}$

$R = 1000 \text{ ft}$

$$L_S = 1.6 \frac{50^3}{1000} = 200 \text{ ft}$$

The answer is **B**.

4. Airport Design

by Francis McKelvey

The questions relating to airport layout and design can be answered by referring to pages 94 and 95 of the NCEES Handbook, 3rd ed. We will present several of the more common problems encountered when designing an airport.

Example 4.1

The preliminary vertical profile for a transport airport runway centerline is given below. The specifications for the geometric design of the vertical profile of the centerline of a runway are presented on p. 95 of the Handbook.

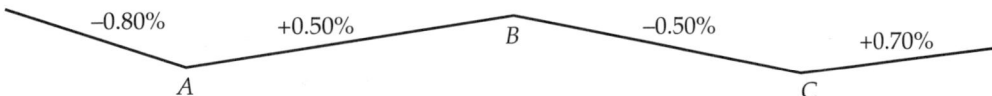

1. The minimum required length of a vertical curve at the point of intersection *A* is most nearly:
   (A) 300 ft
   (B) 600 ft
   (C) 1300 ft
   (D) 2000 ft

2. The minimum required length of a vertical curve at the point of intersection *B* is most nearly:
   (A) 100 ft
   (B) 500 ft
   (C) 1000 ft
   (D) 1500 ft

3. The minimum required length of a vertical curve at the point of intersection *C* is most nearly:
   (A) 200 ft
   (B) 500 ft
   (C) 700 ft
   (D) 1200 ft

4. The minimum required distance between point of intersection A and point of intersection B is most nearly:
   (A) 1000 ft
   (B) 1300 ft
   (C) 2000 ft
   (D) 2300 ft

5. The minimum required distance between point of intersection B and point of intersection C is most nearly:
   (A) 500 ft
   (B) 700 ft
   (C) 2200 ft
   (D) 2500 ft

Solutions:

1. **C** p.95 - Transport Airports: $L_1 = 1.30(1000) = 1300$ ft

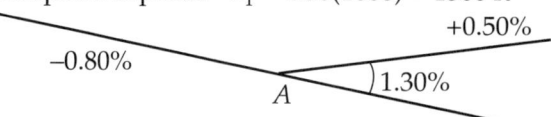

2. **C** p.95 - Transport Airports: $L_2 = 1.00(1000) = 1000$ ft

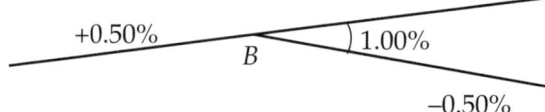

3. **D** p.95 - Transport Airports: $L_3 = 1.20(1000) = 1200$ ft

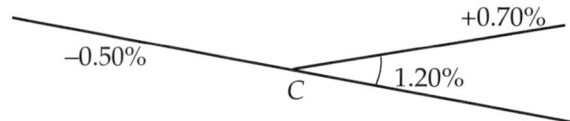

4. **D** p.95 - Transport Airports: $D = 1000(A + B)$
   $= 1000(1.30 + 1.00) = 2300$ ft

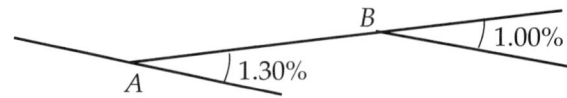

5. **C** p.95 - Transport Airports: $D = 1000(B + C)$
   $= 1000(1.00 + 1.20) = 2200$ ft

Example 4.2

The preliminary vertical profile for an utility airport runway centerline is given. The specifications for the geometric design of the vertical profile of the centerline of a runway are given on Page 95.

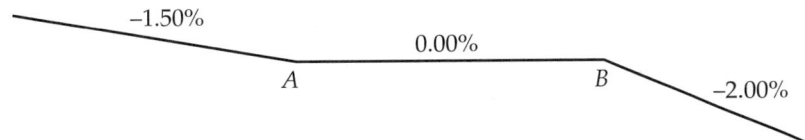

1. The minimum required length of a vertical curve at the point of intersection A is most nearly:
   (A) 150 ft
   (B) 300 ft
   (C) 450 ft
   (D) 600 ft

2. The minimum required length of a vertical curve at the point of intersection B is most nearly:
   (A) 100 ft
   (B) 200 ft
   (C) 300 ft
   (D) 600 ft

3. The minimum required distance between point of intersection A and point of intersection B is most nearly:
   (A) 150 ft
   (B) 200 ft
   (C) 675 ft
   (D) 875 ft

4. The maximum gradient of the runway centerline between point of intersection A and point of intersection B is most nearly:
   (A) 0.00%
   (B) +0.50%
   (C) +1.00%
   (D) +1.50%

Solutions:

1. **C** p.95 - Utility Airport: $L_1 = 300(1.50) = 450$ ft

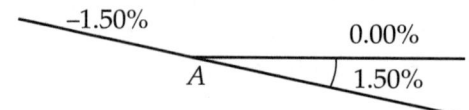

2. **D** p.95 - Utility Airports: $l_2 = 300(2.00) = 600$ ft

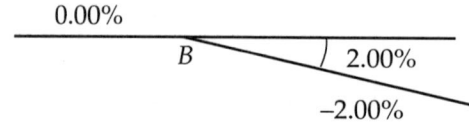

3. **D** p.95 - Utility Airports: $D = 250(A + B)$

$$= 250(1.50 + 2.00) = 875 \text{ ft}$$

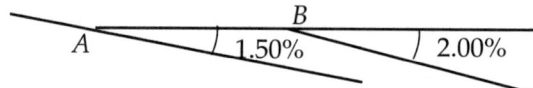

4. **A** From p.95 the maximum gradient = 2.00%. But also the maximum grade change = 2.00%.

$A + A$:

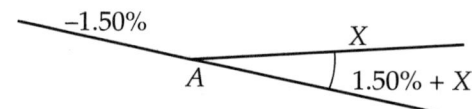

$1.50 + X \leq 2.00$. $\therefore X = 0.5\%$

$A + B$:

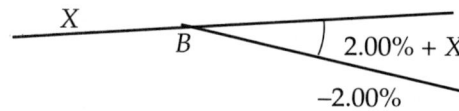

$X + 2.00 \leq 2.00$. $\therefore X = 0.00\%$

Example 4.3

The historical pattern of wind at an airport location has been plotted in a windrose given below. The wind speeds plotted are in the ranges between 0-4, 4-15, 15-31 and 31-47 miles per hour. The airport runway is to be designed for aircraft weighing more than 12,500 pounds. The specifications for wind analysis are given on Page 94 of the NCEES Handbook.

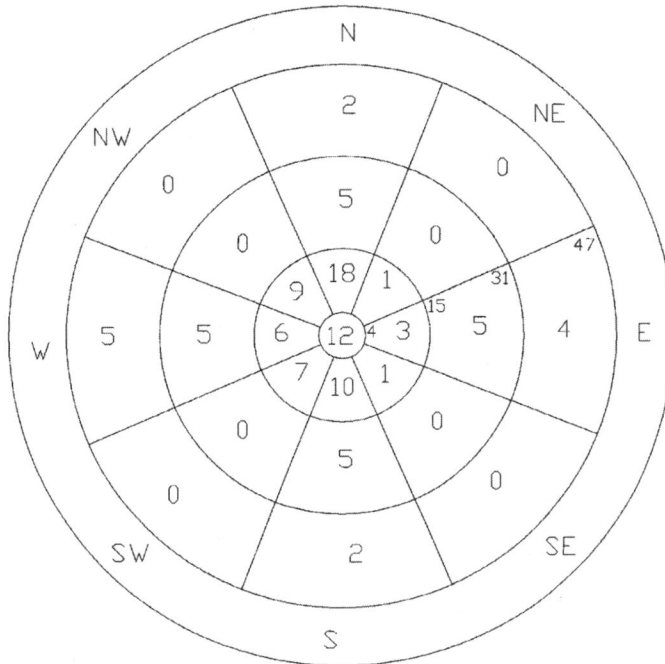

1. The optimal runway orientation at this airport is most nearly:
   (A) N-S
   (B) NW-SE
   (C) E-W
   (D) NE-SW

2. The percentage of time a runway in the N-S direction may be used by aircraft weighing more than 12,500 pounds at this airport without exceeding cross-wind specifications is most nearly:
   (A) 67%
   (B) 81%
   (C) 87%
   (D) 95%

3. The percentage of time the winds come from the NW and have a wind speed of between 4 and 15 miles per hour is:
   (A) 1%
   (B) 9%
   (C) 13%
   (D) 21%

80   Ch. 4 / Airport Design

4. The percentage of time a runway in the E-W direction may be used by aircraft taking off and landing toward the East without exceeding cross-wind specifications is most nearly:
   (A) 28%
   (B) 40%
   (C) 56%
   (D) 68%

Solutions:

1. **C**   Based on wind data given on the next four pages, the runway orientation and percent uses are:

   N-S: 81%

   NW-SE: 77%

   E-W: 86%

   NE-SW: 77%

   Therefore, E-W maximized wind coverage.

1. **Continued** ---- Runway in N-S direction ----

   p.94 infers the maximum cross wind to be 15 mph.

   % = 12+18+1+3+1+10+7+6+9+5+5+2+2 = 81%

   The N-S runway may be used 81% of the time with cross winds less than 15 mph.

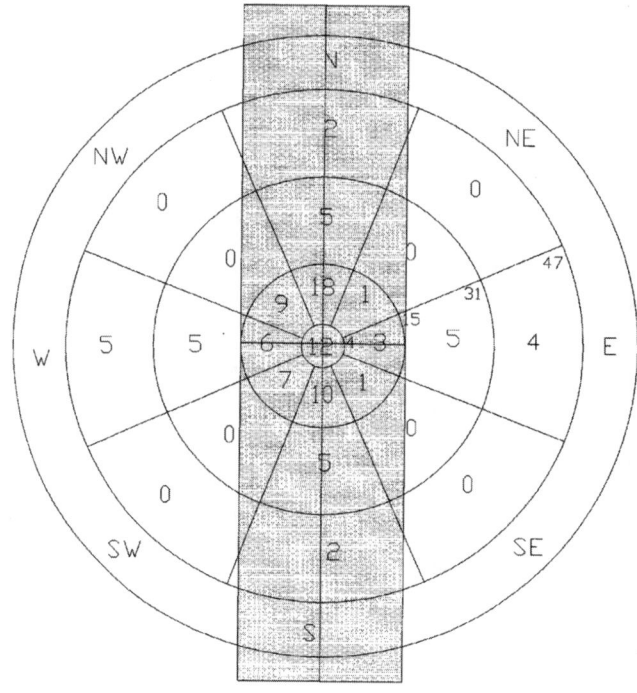

1. **Continued** ---- Runway in the NW-SE direction ----

   p.94 infers the maximum cross wind to be 15 mph.

   % = 12+18+1+3+1+10+7+6+9+2.5+2.5+2.5+2.5 = 77%

   The NW-SE runway may be used 77% of the time with cross winds less than 15 mph.

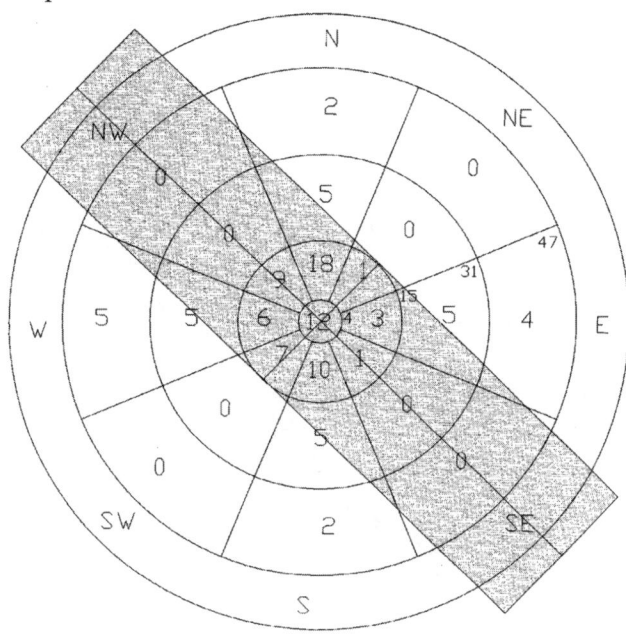

1. **Continued**---- Runway in E-W direction ----

   p.94 infers the maximum cross wind to be 15 mph.

   % = 12+18+1+3+1+10+7+6+9+5+5+5+4 = 86%

   The E-W runway may be used 86% of the time with cross winds less than 15 mph.

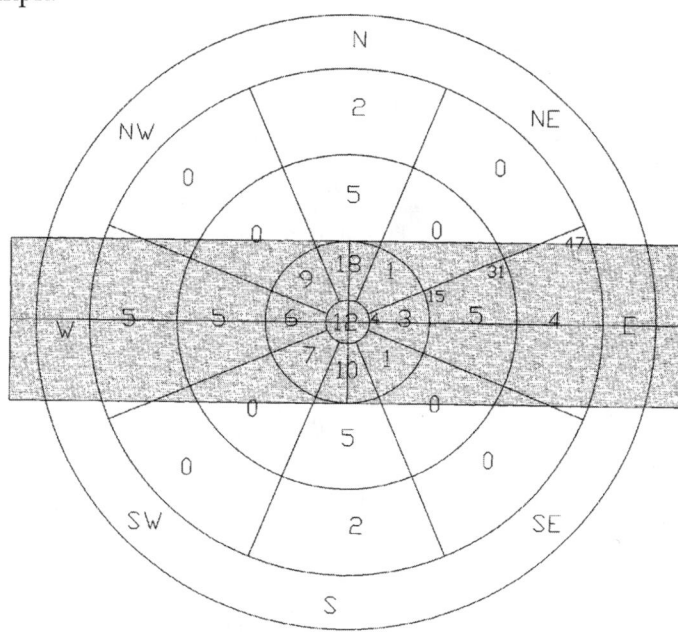

1. **Continued** ---- Runway in NE-SW direction:

   p.94 infers the maximum cross wind to be 15 mph.

   % = 12+18+1+3+1+10+7+6+9+2.5+2.5+2.5+2.5 = 77%

   The NE-SW runway may be used 77% of the time with cross winds less than 15 mph.

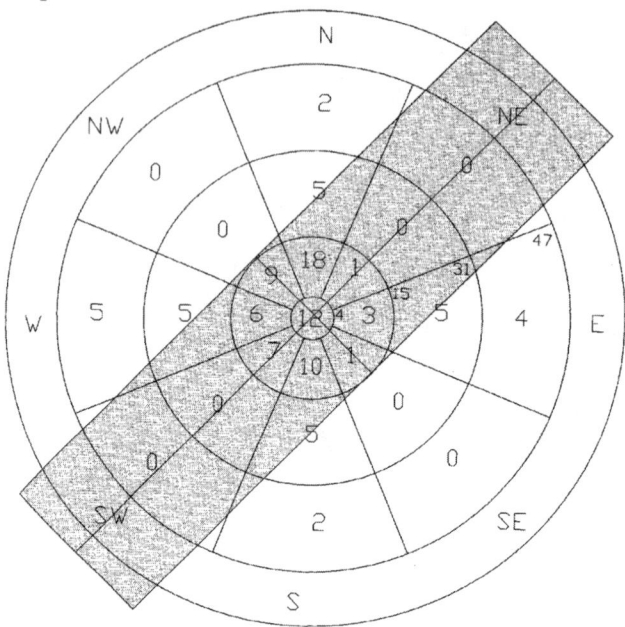

2. **B**   For a runway in the N-S direction:

   % = 12+18+1+3+1+10+7+6+9+5+5+2+2 = 81%

   The N-S runway may be used 81% of the time with cross winds less than 15 mph.

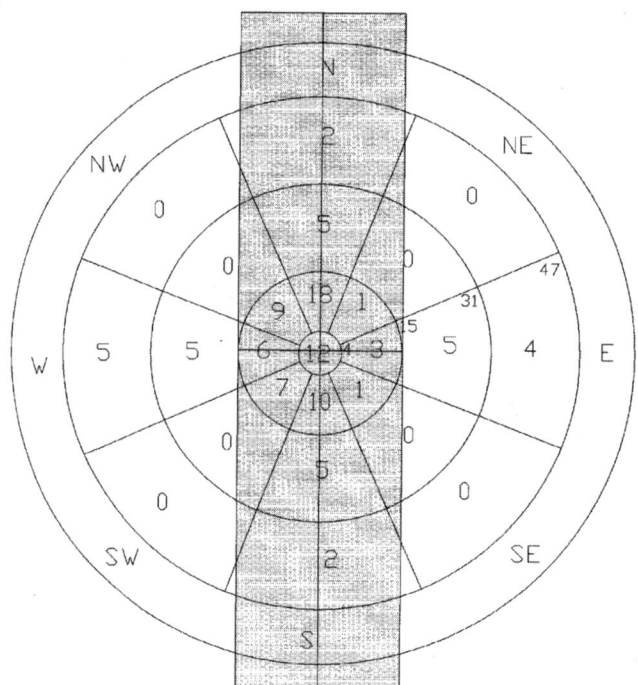

3. **B** The shaded area represents the percentage of time the wind comes from the NW and is between 4 and 15 mph. Therefore, 9% of the time.

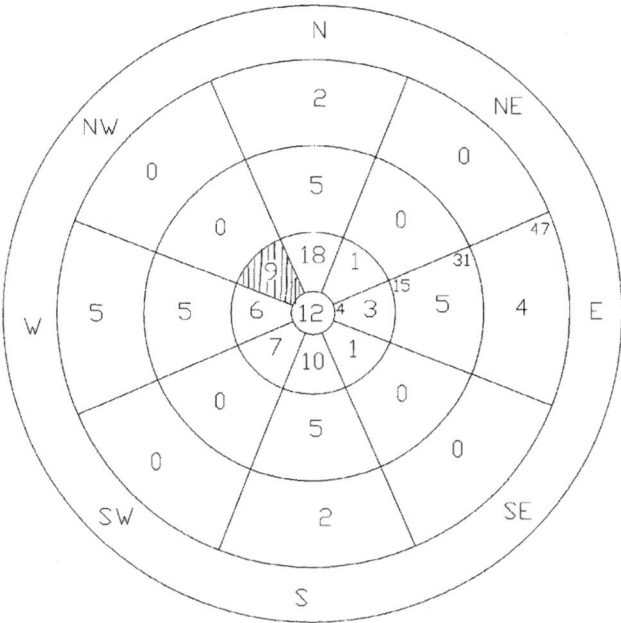

4. **B** The shaded area represents the percentage of time a runway may be used in an East direction so that cross winds are less than 15 mph:

    % = 12+9+1+3+1+5+5+4 = 40% of the time.

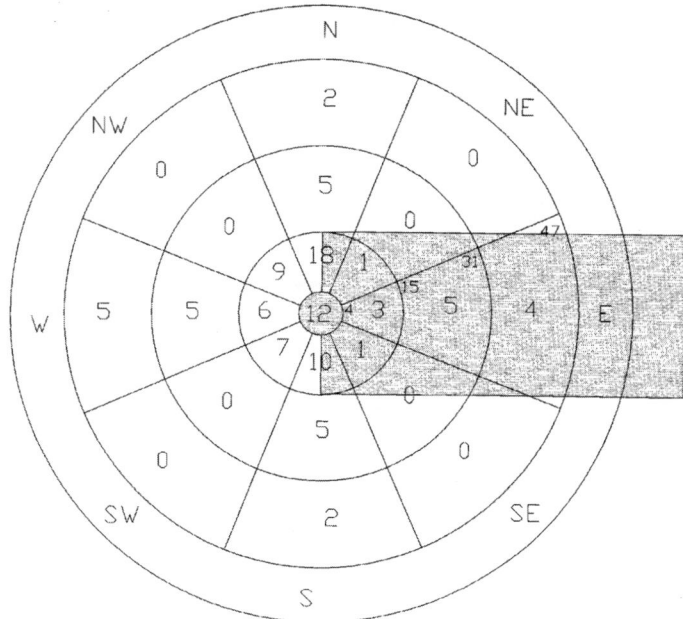

5. Pavement Design

by Gilbert Baladi

Problems involving pavement design can be worked referring to the equations and table on page 95 of the NCEES Handbook, 3rd ed. The following examples illustrate the solution technique.

Example 5.1

A coal fired power plant is served by 25 trucks. When loaded, each truck consists of one single-axle of 14,000 pounds (the front axle), two tandem-axles of 32,000 pounds each (the middle axles), and one single-axle of 22,000 pounds (the back axle). Each truck makes 25 daily trips. The weights on the same axles when the truck is empty and leaving the power plant are 10,000, 12,000, and 8,000. A two-lane road is to be designed to serve the power plant and to last for a period of 10 years.

1. What is the load equivalency factor for a 14,000-pound single-axle?
   (A) 0.027
   (B) 0.36
   (C) 0.189
   (D) 1.0

2. What is the load equivalency factor for a 32,000-pound tandem-axle?
   (A) 8.88
   (B) 6.97
   (C) 0.36
   (D) 0.857

3. What is the design ESAL for the incoming traffic?
   (A) 9,704,438
   (B) 388,176
   (C) 970,444
   (D) 26,588

4. What is the design ESAL for the outgoing traffic?
   (A) 344,012
   (B) 943
   (C) 34,402
   (D) 1,376

Solutions:

1. **B** From the table on page 95, the load equivalency factor for a 14,000-pound single-axle load is 0.36.

2. **D** From the table on page 95, the load equivalency factor for a 32,000-pound tandem-axle load is 0.857.

3. **A** To calculate the design ESAL for the incoming traffic, first calculate the ESAL per truck. This implies that the load equivalency factor (LEF) for each axle on the truck must be obtained:

   For the 14,000-pound single-axle, LEF = 0.36.
   For the 32,000-pound tandem-axle, LEF = 0.857.
   For the 220,000-pound single-axle, LEF = 2.18.

   The total ESAL per truck = 1(0.36) + 2(0.857) + 1(2.18) = 4.254 ESAL/truck.

   For 25 trucks, with each truck making 25 trips per day, the total ESAL per day is
   $$25(25)(4.254) = 2658.75 \text{ ESAL/day}$$
   For a 10-year design life, the total design ESAL is
   $$10(365)(2658.75) = 9{,}704{,}438 \text{ ESAL}$$

4. **A** To calculate the design ESAL for the outgoing traffic, follow the same steps as in the solution to question 3, except use the axle weights of the empty truck:

   For the 10,000-pound single-axle, LEF = 0.0877.
   For the 12,000-pound tandem-axle, LEF = 0.0144.
   For the 8,000-pound single-axle, LEF = 0.0343.

   The total ESAL per truck = 1(0.0877) + 2(0.0144) + 1(0.0343) = 0.1508 ESAL/truck.

   For 25 trucks, with each truck making 25 trips per day, the total ESAL per day is
   $$25(25)(0.1508) = 94.25 \text{ ESAL/day}$$
   For a 10-year design life, the total design ESAL is
   $$10(365)(94.25) = 344{,}013 \text{ ESAL}$$

Example 5.2

For the power plant of Example 5.1, the design structural number for the incoming traffic (loaded trucks) is 5.2 and for the outgoing traffic (empty trucks) is 3.0. The design engineer used the following material for road construction:

   A 12-inch-thick sand sub-base with a layer coefficient of 0.10
   A 6-inch-gravel base with a layer coefficient of 0.13
   An asphalt-concrete surface layer with a layer coefficient of 0.40

1. What is the required asphalt-concrete thickness for the incoming traffic (loaded trucks)?
   (A) 8 in.
   (B) 6 in.
   (C) 10 in.
   (D) 4 in.

2. What is the required asphalt-concrete thickness for the outgoing traffic (empty truck)?
   (A) 3 in.
   (B) 3.6 in.
   (C) 2.6 in.
   (D) 4 in.

3. For the material data given above, what is the required gravel thickness that will produce the same structural number as a one-inch-thick asphalt-concrete layer?
   (A) 3.1 in.
   (B) 0.33 in.
   (C) 6 in.
   (D) 1 in.

Solutions:

1. **A** Use the equation on page 95:
   $$SN = a_1 D_1 + a_2 D_2 + a_3 D_3$$
   Solving for the thickness of the asphalt-concrete yields
   $$D_1 = \frac{SN - a_2 D_2 - a_3 D_3}{a_1}$$
   Therefore,
   $$D_1 = \frac{5.2 - 0.1(12) - 0.13(6)}{.4} = 8 \text{ in.}$$

2. **C** Solving for D_1 by following the same steps as in the solution to Question 1 yields
   $$D_1 = \frac{3.0 - 0.1(12) - 0.13(6)}{.4} = 2.6 \text{ in.}$$

3. **A** The structural number (SN) of a one-inch asphalt-concrete layer is $1(0.4) = 0.4$. The required thickness D_2 of the gravel that will yield a structural number of 0.4 can be calculated from the following equation:
   $$SN_2 = a_2 D_2$$
   Solving for D_2 gives
   $$D_2 = SN_2 / a_2$$
   $$= 0.4 / 0.13 = 3.1 \text{ in.}$$

Example 5.3

Two single-axle trucks travel one road in one direction twice each day. The first truck consists of a 10,000-pound front single-axle and an 18,000-pound rear single-axle. The second truck consists of a 12,000-pound front single-axle and a 24,000-pound rear single-axle. Assume that, for each trip, the first truck delivers a unit of damage to the road.

1. For each trip, how many units of damage are delivered to the road by the second truck?
   (A) 3.22
   (B) 2.96
   (C) 1.09
   (D) 0.34

2. For each day, how many units of damage are delivered to the road by the second truck?
   (A) 6.44
   (B) 2.18
   (C) 4.62
   (D) 5.92

Solutions:

1. **B** The relative damage delivered by each truck is proportional to the ESAL of each truck. Thus, the ESAL of each truck must be calculated first.

   Truck 1:

   > The LEF for a 10,000-pound single-axle = 0.0877
   > The LEF for an 18,000-pound single-axle = 1.0
   > The total ESAL of truck 1 = 1.0 + 0.0877 = 1.0877

   Truck 2:

   > The LEF for a 12,000-pound single-axle = 0.189
   > The LEF for a 24,000-pound single-axle = 3.03
   > The total ESAL of truck 2 = 3.03 + 0.189 = 3.219

   The number of units of damage delivered to the pavement by truck 2 are
   $$3.219 / 1.0877 = 2.96$$

2. **D** The second truck makes two trips per day. The number of units of damage are
   $$2(2.96) = 5.92$$

Example 5.4

The required structural number SN of an asphalt pavement to protect the roadbed soil is 3.5. The required structural number to protect the sub-base material is 2.84. The required structural number to protect the base material is 2.0. The pavement is constructed of the following materials:

    Sand sub-base with a layer coefficient of 0.11.
    Gravel base with a layer coefficient of 0.14.
    Asphalt-concrete surface with a layer coefficient of 0.4.

1. What is the required thickness of the asphalt-concrete layer?
   (A) 6.0 in.
   (B) 5.0 in.
   (C) 4.0 in.
   (D) 3.0 in.

2. What is the required thickness of the base layer?
   (A) 6.0 in.
   (B) 5.0 in.
   (C) 4.0 in.
   (D) 3.0 in.

3. What is the required thickness of the sub-base layer?
   (A) 6.0 in.
   (B) 5.0 in.
   (C) 4.0 in.
   (D) 3.0 in.

Solutions:

1. **B**   The AASHTO structural number SN equation is given as
$$SN = a_1 D_1 + a_2 D_2 + a_3 D_3$$
where the a_1 and D_1 are the layer coefficient and layer thickness. The above equation can be rewritten as follows:
$$SN = SN_1 + SN_2 + SN_3$$
where SN is the structural number of the pavement, and SN_i is the SN of layer i. Since the required SN to protect the base material is 2.0, it is inferred that the SN of the asphalt-concrete is 2.0. Therefore:
$$SN_1 = a_1 D_1 = 2.0$$
We solve for D_1:
$$D_1 = SN_1 / a_1 = 2.0 / 0.4 = 5 \text{ in.}$$

2. **A**   Given that the required structural number to protect the sub-base is 2.84, we have

$$SN = a_1 D_1 + a_2 D_2 = 2.84$$

Since $a_1 = 0.4$, $D_1 = 5$ in., and $a_2 = 0.14$, we solve for D_2 as follows:

$$D_2 = \frac{SN - a_1 D_1}{a_2}$$

Therefore,

$$D_2 = \frac{2.84 - 0.4 \times 5}{0.14} = 6 \text{ in.}$$

3. **A** The required thickness of the sub-base layer can be obtained by solving the following structural number equation:

$$SN = a_1 D_1 + a_2 D_2 + a_3 D_3$$

All variables are known except the thickness D_3 of the sub-base. Solving for D_3:

$$D_3 = \frac{SN - a_1 D_1 - a_2 D_2}{a_3}$$

Therefore,

$$D_3 = \frac{3.5 - 0.4 \times 5 - 0.14 \times 6}{0.11} = 6 \text{ in.}$$

6. Environmental Engineering

by Susan J. Masten

To work problems relating to Environmental Engineering, it will be necessary to refer to the equations on page 92 and the figures on page 98 of the NCEES Handbook, 3rd ed. It may also be necessary to refer to the information contained in the Chemistry section on pages 61-63.

6.1 Hardness Removal

Many groundwaters (and a few surface waters) contain excessive minerals which must be removed prior to distribution of the treated water. Hardness, which is caused by the presence of multivalent cations, usually calcium and magnesium, is problematic because of (1) the formation of slimy, sticky deposits as a result of the interactions between the hardness ions and the soap, (2) the tendency of soaps not to produce a good lather, (3) the formation of hard, rock-like deposits in pipes and boilers.

Hardness may be present as either carbonate or noncarbonate hardness. Carbonate hardness is that portion of the (total) hardness that is associated with carbonate and bicarbonate anions. It is easily removed by precipitation (as calcium carbonate) upon heating of the water. As such, carbonate hardness causes scaling problems in water heaters and hot water pipes. Noncarbonate hardness is that portion of the hardness minerals that is associated with anions other than carbonate and bicarbonate. This form of hardness is more difficult to remove and does not precipitate as the temperature of the water is increased.

The removal of hardness minerals can be accomplished by softening or by ion exchange. Softening is usually accomplished with lime. If insufficient alkalinity is present, then soda ash is also used. Following softening, the pH of the treated water must be reduced to approximately 9.0. This is accomplished using recarbonation, in which carbon dioxide is bubbled into the water. The carbon dioxide dissolves and is hydrolyzed to form carbonic acid which lowers the pH. Ion exchange results in the exchange of Ca^{2+} and Mg^{2+} ions with equivalent number of sodium (Na^{2+}) ions. For example, calcium is removed by the reaction:

$$Ca(HCO_3)_2 + Na_2R \rightarrow CaR + 2NaHCO_3$$

where R is the resin support. Refer to the lime-soda softening equations (p. 92) along with the table of equivalent weights, molecular weights and equivalents weights (p. 92) in the NCEES Handbook to solve these problems.

Example 6.1

A large university serving 44,000 students and 5,000 faculty and staff obtains its drinking water from an aquifer. Of the 44,000 students, half live on campus. The water usage is:

    Resident students (those living on campus): 100 gal/capita/day
    Nonresident students, faculty and staff: 50 gal/capita/day

A partial analysis of the water has shown that it has the following characteristics (given as mg/L as $CaCO_3$, except for pH):

pH	8.1	Alkalinity	280.0
Ca^{2+}	365.0	Mg^{2+}	74.0
Fe^{3+}	2.3	Na^+	50.0
K^+	26.0	SO_4^{2-}	236.0
CO_2	47.0		

A water has a calcium concentration of 20.0 mg/L (as the ion) after softening. The pH of the softened water is 10.8.

1. Calculate the theoretical amount of lime (in kg/day) that would be required to soften this water. Note that to raise the pH to precipitate magnesium hydroxide, an excess of 1.25 meq/L of lime must be provided.
   (A) 3,987
   (B) 3,251
   (C) 4,608
   (D) 2,784

2. Calculate the amount of calcium carbonate ($CaCO_3$) sludge (in kg) produced each day by the reaction of carbon dioxide with lime.
   (A) 167
   (B) 334
   (C) 668
   (D) 513

Solutions

1. **C** We must first calculate the concentrations of each of the reactive species using the data provided in the table of equivalent weights given on page 92 of the NCEES Handbook:

$$[CO_2] = \frac{47.0}{50.04} = 0.94 \text{ meq/L}$$

$$[Ca^{2+}] = \frac{365.0}{50.04} = 7.29 \text{ meq/L}$$

$$[Mg^{2+}] = \frac{74.0}{50.04} = 1.48 \text{ meq/L}$$

$$[HCO_3^-] = \frac{280.0}{50.04} = 5.60 \text{ meq/L}$$

(Note: As the pH = 8.1, Alk $\cong [HCO_3^-]$)

We must also calculate the flow rate for the treatment plant:

$$Q = (22{,}000 \times 100 \text{ gal/cap/day}) + [(22{,}000 + 5000) \times$$
$$50 \text{ gal/cap/day}] = 3.55 \text{ MGD}$$

The amount of lime required can be calculated using equations 1, 2, and 4 (Lime-Soda Softening Equations) on page 92 of the NCEES Handbook.

	meq/L	meq/L lime required	meq/L soda ash required
CO_2	0.94	0.94	
$Ca(HCO_3)$	5.60	5.60	
$CaSO_4$	1.69	—	1.69
$MgSO_4$	1.48	1.48	1.48
excess		1.25	
		9.27	

Mass of lime per day =

$(9.27 \text{ meq Ca(OH)}_2/L) \times (37.046 \text{ mg Ca(OH)}_2/\text{meq}) \times (3.785 \text{ L/gal})$

$\times (3.55 \times 10^6 \text{ gal/day}) \times (\text{kg}/10^6 \text{ mg})$

= 4608 kg/day

2. **A** We can calculate the amount of calcium carbonate sludge produced using the stoichiometric equation given on page 92 of the NCEES Handbook:

$$CO_2 + Ca(OH)_2 \rightarrow CaCO_{3(s)} + H_2O$$

The concentration of carbon dioxide in the water is 0.47 mmol/L (= 47.0 mg/L as $CaCO_3/100.086$ mg/mmol). Since 1 mole of carbon dioxide reacts to form one mole of calcium carbonate, 0.47 mmol/L of calcium carbonate is produced.

Mass of sludge produced =

$(0.47 \text{ mmol/L}) \times (100.08 \text{ mg/mmol}) \times$

$(3.55 \times 10^6 \text{ gal/d}) \times (1 \text{ kg}/10^6 \text{ mg})$

= 167.0 kg/d

6.2 Grit Chamber Design

The initial steps in municipal wastewater treatment involve pretreatment using screens, comminutors, grit chambers and/or equalization basins.

Bar screens or racks are used to remove large objects like tree limbs and small animals that may get into the wastewater plant influent through storm sewer entrances or displaced sewer access (manhole) covers. These materials can seriously affect pumps and piping within the plant. Bar screens or racks may be mechanically or manually cleaned. Generally only smaller facilities have manually screened racks. Larger plants may use communitors in place of bar screens. Communitors grind up or chop large objects so that they do not clog pipes or pumps but can be removed in downstream processes.

Grit chambers follow bar screens and are used to remove grit, which consists primarily of sand, cinders, coffee grounds and pea gravel. Grit can originate from storm runoff or from material disposed of in home or commercial garbage disposal units.

Equalization basins are used to dampen diurnal variations in flow, BOD and suspended solids loading rates. They also enhance the degree of treatment in downstream processes. These basins are usually constructed of concrete, although earth or steel may be used. The wastewater in the basin is aerated to provide mixing, to keep the solids in suspension and to prevent the wastewater from becoming septic (anoxic).

The approach velocity is used in designing the first three of these unit operations. Use the equations given on p. 92 of the NCEES Handbook to design a grit chamber.

Example 6.2

A municipal wastewater is pretreated using a horizontal flow grit chamber having the following dimensions:

  Length = 60'
  Width = 4'
  Depth = 6'

The design flow rate is 10.83 cfs. Calculate the approach velocity in the chamber.
  (A) 0.05 ft/s
  (B) 0.03 ft/s
  (C) 0.68 ft/s
  (D) 0.45 ft/s

Solution:

A_x = Cross-sectional area in grit chamber = $4' \times 6' = 24 \text{ ft}^2$. Using the equation for approach velocity given on page 92 of the NCEES Handbook, the approach velocity can be calculated as follows:

$$\text{approach velocity} = \frac{Q}{A_x} = \frac{10.83 \text{ ft}^3/\text{s}}{24 \text{ ft}^2} = 0.45 \text{ ft/s}$$

The answer is **D**.

6.3 Primary Settling

Primary treatment of municipal wastewater involves gravity settling of discrete particles. In some cases, coagulant aids may be added to enhance settling. The main uses of primary settling in wastewater treatment are the removal of settleable solids and floating material such as oil and grease. Primary settling basins can achieve from 50 to 70% removal of suspended solids and 25 to 40% removal of BOD_5. These basins usually have a detention time from 1.5 to 3 h. Primary sedimentation basins may be rectangular, square, or circular in shape.

The detention time and weir loading are important design parameters for sedimentation basins. Detention time is also used in designing any type of process which involves chemical addition, including rapid mix, flocculation, chlorination and ozonation. Overflow (hydraulic loading) rates are commonly used for designing both sedimentation basins and filtration units.

Example 6.3

A municipal wastewater having a design flow rate of 2.5 MGD is to be treated using 2 primary clarifiers. The clarifiers each have the following characteristics:

$$\text{Diameter} = 43'$$
$$\text{Depth} = 10'$$
$$\text{Weir length} = 270'$$

1. Calculate the detention time (hydraulic residence time) in each of the clarifiers.
   (A) 4.16 hr
   (B) 2.08 hr
   (C) 0.52 hr
   (D) 1.04 hr

2. Calculate hydraulic loading (overflow) rate in units of $\text{ft}^3/\text{ft}^2\text{-hr}$.
   (A) 2.40
   (B) 4.79
   (C) 9.59
   (D) 19.18

3. What is the weir loading rate in units of $ft^3/ft\text{-}hr$?
   (A) 25.8
   (B) 12.9
   (C) 6.45
   (D) 51.6

Solutions:

1. **B** Use the equation for detention time (t_d) given on page 92 of the NCEES Handbook. Since the total flow is divided between two clarifiers, each clarifier is to be designed based upon a flow rate of 1.25 MGD:

   $$Q = (1.25 \times 10^6 \text{ gal/day})(0.1337 \text{ ft}^3/\text{gal})(0.417 \text{ day/hr}) = 6962 \text{ ft}^3/\text{hr}$$

   $$t_d = V/Q = \frac{H\pi d^2/4}{6962 \text{ ft}^3/\text{hr}} = \frac{10 \text{ ft } \pi (43 \text{ ft})^2/4}{6962 \text{ ft}^3/\text{hr}} = 2.08 \text{ hr}$$

   where d = diameter and H = depth of the clarifier.

2. **B** The equation for the overflow rate is

   $$\text{Overflow rate} = Q/A = \frac{6962 \text{ ft}^3/\text{hr}}{\pi d^2/4} = 4.79 \text{ ft}^3/\text{ft}^2\text{-hr}$$

3. **A** The equation for the weir loading rate provides

   $$\text{Weir loading rate} = Q/L = \frac{6962 \text{ ft}^3/\text{hr}}{270 \text{ ft}} = 25.8 \text{ ft}^3/\text{ft-hr}$$

6.4 Activated Sludge

Activated sludge treatment uses a consortium of microorganisms in a well-mixed tank to oxidize organic matter under aerobic conditions. The main purpose of activated sludge treatment is to remove soluble BOD from the wastewater. The main components of the system include a biological reaction basin (aeration basin), a clarifier for sedimentation of the biomass, and recycle sludge pumps to recycle a portion of the biomass (return activated sludge, RAS) back to the head of the aeration basin. The activated sludge-wastewater mixture is referred to as mixed liquor suspended solids (MLSS). Typically, the concentration of MLSS ranges from 2,000 to 4,000 mg/L. The concentration of microorganisms is often quantified as mixed liquor volatile suspended solids (MLVSS). Because these systems use aerobic organisms to consume the organic matter, either air or pure oxygen must be supplied in the aeration basins. This is usually accomplished by either diffused aerators or by mechanical aerators, the former being more efficient.

Among the major parameters used in designing activated sludge systems are: F:M (organic loading rate), the volumetric organic loading rate, the solids retention time and the solids loading rate to the secondary clarifier.

Example 6.4

A municipal wastewater treatment plant is to be designed. The design flow rate is 2.5 MGD. The wastewater has the following characteristics after primary treatment:

> Influent BOD = 200 mg/L
> Influent Suspended Solids = 200 mg/L

The wastewater must have the following characteristics after secondary treatment.

> Effluent BOD = 15 mg/L
> Effluent SS = 20 mg/L

A preliminary design for a complete-mix activated sludge system has yielded the following design parameters:

> Sludge volume index (SVI) = 125
> Return activated sludge concentration = 8000 mg/L
> MLSS = 3500 mg/L
> Recycle = 50 %
> Depth = 15 ft
> Width = 50 ft
> Length = 100 ft

Sedimentation of the waste activated sludge is accomplished using secondary clarification. Two secondary clarifiers (each 28 ft. in diameter) are proposed.

1. Calculate *F:M* in units of mg BOD applied/mg MLSS-day.
   - (A) 0.255
   - (B) 0.128
   - (C) 0.111
   - (D) 0.510

2. Calculate the volumetric organic loading rate in units of kg BOD/m^3-day.
   - (A) 0.45
   - (B) 15.6
   - (C) 35.6
   - (D) 0.89

3. Calculate the solids residence time.
   - (A) 10.7 hr
   - (B) 4.7 hr
   - (C) 0.44 hr
   - (D) 1.01 hr

4. Calculate the solids loading rate in lb/ft²-day to the secondary clarifier.
   (A) 135
   (B) 118.4
   (C) 59.2
   (D) 270

Solutions:

1. **A** $V_A = (15 \text{ ft})(50 \text{ ft})(100 \text{ ft}) = 75{,}000 \text{ ft}^3$

   The equation for the F:M ratio can be obtained from page 92 of the NCEES Handbook.

   $$F{:}M = \frac{QS_0}{V_A X_A} = \frac{2.5 \times 10^6 \text{ gal/day}(0.1337 \text{ ft}^3/\text{gal})(200 \text{ mg/L})}{75{,}000 \text{ ft}^3 (3{,}500 \text{ mg/L})}$$

   $= 0.255$ mg BOD applied/mg MLSS-day

2. **D** The equation for the volumetric organic loading rate is:

   Volumetric organic loading rate $= QS_0 / V =$

   $$\frac{2.5 \times 10^6 \text{ gal/day}(200 \text{ mg/L})(0.1337 \text{ ft}^3/\text{gal})(10^3 \text{ L/m}^3)(1 \text{ kg}/10^6 \text{ mg})}{75{,}000 \text{ ft}^3}$$

   $= 0.89$ kg BOD/m³-day

3. **B** The equation for the solids residence time is:

   $$\text{Solids residence time} = \frac{V_A X_A}{(Q_w X_w + Q_e X_e)} =$$

   $$\frac{75{,}000 \text{ ft}^3 (3500 \text{ mg/L})}{\{0.5(2.5 \times 10^6 \text{ gal/day})(8000 \text{ mg/L}) + 2.5 \times 10^6 \text{ gal/day}(20 \text{ mg/L})\} 0.1337 \text{ ft}^3/\text{gal}}$$

   $= 0.19$ day $= 4.7$ hr

4. **C** The equation for the solids residence time can be obtained from page 92 of the NCEES Handbook.

   Solids loading rate $= QX/A =$

   $$\frac{1.25 \times 10^6 \text{ gal/day}(3.785 \text{ L/gal})(3500 \text{ mg/L})(1 \text{ g}/10^3 \text{ mg})(1 \text{ lb}/454 \text{ g})}{\pi (28 \text{ ft})^2 / 4}$$

   $= 59.2$ lb/ft²-day

6.5 Fixed-film Processes

Tricking filters and rotating biological contactors (RBCs), i.e., attached growth processes, use a consortium of microorganisms that are attached to the surface of a medium to oxidize BOD in wastewater. For this reason, they are referred to as fixed-film or attached-growth processes. The medium used in trickling filters is stationary, while the disks used in RBCs rotate around a central axis in a basin containing the wastewater. As with activated sludge treatment, the purpose of these processes is to reduce the concentration of soluble BOD in the wastewater. As the wastewater contacts the microorganisms in the biofilm on the trickling filter medium (usually rock or plastic) or on the RBC plates, the microorganisms use the oxygen demanding material in the wastewater as a growth substrate. As with activated sludge, oxygen must be provided for these processes to be effective.

Example 6.5

The wastewater flow rate from the City of Oz is 1.20 MGD. The influent BOD_5 and suspended solids (SS) are 240 mg/L and 220 mg/L, respectively. The wastewater is treated by primary sedimentation which removes 33% of the BOD_5 and 35% of the SS followed by a 4-stage RBC unit. The effluent BOD_5 must be less than 20 mg/L. The total surface area of the discs is 700,000 ft^2. Calculate the organic loading rate (based on the surface area) in units of lb BOD/10^3 ft^2-day to the RBC unit.

(A) 4.2
(B) 1.41
(C) 2.30
(D) 0.53

Solution:

The equation for the organic loading rate can be obtained from page 92 of the NCEES Handbook:

$$\text{Organic loading rate} = \frac{QS_o}{A_m} =$$

$$\frac{1.20 \times 10^6 \text{ gal/day}(240 \text{ mg/L})(0.67)(3.785 \text{ L/gal})(1 \text{ mg}/10^3 \text{ g})(\text{lb}/454 \text{ g})}{700,000 \text{ ft}^3}$$

$$= 2.30 \text{ lb BOD}/10^3 \text{ ft}^2\text{-day}$$

The answer is **C**.

6.6 Sludge Production

Sludge is produced by a number of processes. For example, in water treatment, chemical sludges are produced by the coagulation of naturally occurring organic matter and turbidity in surface waters and by the softening of groundwaters. Sludges are also produced during primary and secondary clarification of municipal wastewater. Since sludge contains predominantly water, sludge processing units must be designed based upon the wet weight of the sludge, rather than the dry weight.

Example 6.6

The primary sludge from the City of Oz (see Example 6.5) is 3.5% solids. The density of the sludge is 1020 kg/m^3.

1. Calculate the sludge production rate on a dry weight basis.
   (A) 1,000 kg/day
   (B) 382 kg/day
   (C) 35 kg/day
   (D) 1,100 kg/day

2. Calculate the sludge production rate on a wet weight basis.
   (A) 28.0 m^3/day
   (B) 0.98 m^3/day
   (C) 30.8 m^3/day
   (D) 10.7 m^3/day

Solutions:

1. **C** The sludge production rate can be calculated based using the flow rate, the influent SS concentration and the percent removal:

   Sludge production rate =
   $$(1.20 \times 10^6 \text{ gal/day})(220 \text{ mg/L})(0.035)(3.785 \text{ L/gal})(\text{kg}/10^6 \text{ mg})$$
   $$= 35 \text{ kg/day}$$

2. **B** The equation for the sludge production rate on a wet weight basis can be obtained from page 92 of the NCEES Handbook. Note that the parameter, density, in the numerator is the density of the sludge, not the density of water.

   Sludge production rate on a wet weight basis =
   $$\frac{M(100)}{\rho(\% \text{ solids})} = \frac{35 \text{ kg/day} \times 100}{1020 \text{ kg/m}^3 \times 3.5} = 0.98 \text{ m}^3/\text{day}$$

6.7 Chemical Dosing

Numerous chemicals must be added during treatment of water and wastewater. The mass of chemicals that are to be added on a daily basis must be calculated to ensure that a sufficient supply of the required chemicals is maintained on site.

Example 6.7

A water treatment plant having an average daily flow rate of 17.5 MGD adds lime during softening at a dosage of 245.0 mg/L. Calculate the mass of lime that must be supplied each day.

(A) 40,000 lb
(B) 24,000 lb
(C) 35,800 lb
(D) 61,000 lb

Solution:

The mass of a chemical that is to be added can be calculated using the flow rate and the dosing rate. The equation is provided on page 92 of the NCEES Handbook:

$$\frac{\text{mass}}{\text{day}} = \frac{Q \text{ MGD} \times \text{Concentration mg/L} \times 8.34 \text{ lb/MGal}}{\text{mg/L}}$$

$$= 17.5 \text{ MGD} \times 245 \text{ mg/L} \times 8.34 = 35{,}800 \text{ lb/day}$$

The answer is **C**.

6.8 Biochemical Oxygen Demand (BOD)

Biochemical oxygen demand (BOD) is the amount of oxygen required by microorganisms, predominantly bacteria, to oxidize organic matter under aerobic conditions. The BOD test is conducted in the dark (to avoid algae growth) at 20°C and in the presence of an excess of nutrients. This assay (BOD_5) is conventionally run for 5 days. If a wastewater is analyzed for BOD on a daily basis, one would observe that the concentration of BOD increases exponentially and asymptotes toward a constant value, known as the ultimate BOD. The ultimate BOD is usually reached after about 20 days.

Example 6.8

A 5 day BOD was determined to be 250.0 mg/L, while the 20 day BOD (assume that this is the ultimate BOD) was 300.0 mg/L. Determine the BOD rate constant (base e).

(A) 0.36 day^{-1}
(B) 0.037 day^{-1}
(C) 0.16 day^{-1}
(D) 0.016 day^{-1}

Solution:

Use the equation for BOD exertion given on page 92 of the NCEES Handbook to solve this problem:

$$y_t = L(1 - e^{-kt})$$

Solving for k yields:

$$k = -\frac{\ln\left(1 - \frac{y_t}{L}\right)}{t} = -\frac{\ln\left(1 - \frac{250}{300}\right)}{5} = 0.36 \text{ day}^{-1}$$

The answer is **A**.

6.9 Sewage Flow Ratio Curves

Wastewater flows vary depending upon a variety of factors including the season of the year, time of day, affluence of the community, and weather conditions. Wastewater treatment plants are designed based upon low, average and peak dry weather flows. The sewage flow ratio curves (p. 98 of the NCEES Handbook) can be used to estimate the peak and minimum dry weather flows based upon the size of the population served by the treatment plant. These curves are based upon equations and graphical relationships developed from case studies.

Example 6.9

A city of 100,000 people has an average wastewater flow of 10 MGD. The maximum and minimum flows can be estimated by the sewage flow ratio curves. Using the equations $\dfrac{14}{4+\sqrt{P}}$ (Curve B) and $\dfrac{5}{P^{0.107}}$ (Curve A_2) estimate the minimum flow.

(A) 20 MGD
(B) 5 MGD
(C) 4.3 MGD
(D) 3 MGD

Solution:

Using the equation $5/P^{0.107}$ (Curve A_2), we can estimate the minimum flow. The population is 100,000, so read the ratio for the point where the line for 100,000 intersects with curve A_2. The ratio is 0.5. Therefore, the minimum flow is 0.5 × 10 MGD = 5 MGD

The answer is **B**.

6.10 Hydraulic-Elements Graph

Wastewater and storm run-off are commonly collected in circular pipes. In many cases, these sewers do not flow full and it is time-consuming to calculate the hydraulic radii and the cross-sectional area for pipes that are partially full. The Hydraulic-Elements Graph for Circular Sewers (presented on p. 98 of the NCEES Handbook) can be used to calculate partial-flow values from full-flow conditions.

Example 6.10

A circular sewer when flowing full flows at a flow rate of 10 m^3/s and a velocity of 6 m/s.

1. What is the velocity when $Q = 6$ m^3/s?
    (A) 4.2 m/s
    (B) 5.6 m/s
    (C) 6.3 m/s
    (D) 7.2 m/s

2. What is the full-flow depth if the depth of flow when $Q = 6$ m^3/s is 80 cm?
    (A) 2.15 m
    (B) 1.95 m
    (C) 1.65 m
    (D) 1.45 m

Solutions:

Use the chart provided on page 98 of the NCEES Handbook, 3rd ed. The ratio of the flows ($Q/Q_f = 6/10$) is 0.60. Enter this value on the chart and read off the corresponding values for the velocity and depth of flow. Use the dashed lines (for constant Darcy-Weisbach friction factor f and Manning's constant n).

1. **C** Read 1.05 on the horizontal axis using the dashed V-line:
$$V = 1.05 \times 6 = 6.3 \text{ m/s}$$

2. **D** Read 0.55 on the vertical axis. Then
$$\frac{d}{D} = 0.55 \quad \text{or} \quad D = \frac{0.8}{0.55} = 1.45 \text{ m}$$

7. Hydrology and Fluid Flow

by Merle C. Potter

The equation used to solve problems related to hydrology is found on page 94 of the NCEES Handbook, 3rd ed. Other equations on fluid mechanics are also relevant to this subject, particularly those relating to flow rate (discharge or volumetric flow rate) ans mass flux (mass flow rate). Problems involving fluid flows will use the equations found on pages 36 to 44 of the Handbook. Also, review the problems in the chapter on Fluid Mechanics in the Mechanical Engineering part of this book, especially those in the sections on the energy equation and on fans, pumps and turbines.

Example 7.1

The runoff from a 50-hectare parcel, in the Chicago area, flows through a 180-meter-long storm sewer pipe. The pipe is designed for a 10-year, 30-minute duration storm. The ground elevation of the pipe inlet is 201 m and it is 200 m for the pipe outlet. The dimensionless runoff coefficient is 0.3 and the Manning n is 0.015. Two intensity-duration curves for the Chicago area follow.

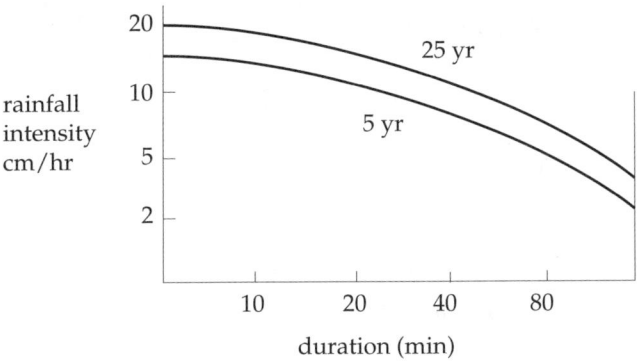

1. Estimate the flow rate.
   (A) $4 \text{ m}^3/\text{s}$
   (B) $3 \text{ m}^3/\text{s}$
   (C) $2 \text{ m}^3/\text{s}$
   (D) $1 \text{ m}^3/\text{s}$

2. The slope of the sewer pipe is:
   (A) 0.0078
   (B) 0.0067
   (C) 0.0056
   (D) 0.0045

3. The diameter of the sewer pipe should be sized at:
   (A) 4 m
   (B) 3 m
   (C) 2 m
   (D) 1.5 m

Solutions:

1. **A** Use the formula at the top of p. 94:
$$Q = ciA = 0.3 \times 10 \times 50 \times (10^4 \times \frac{1}{100} \times \frac{1}{3600}) = 4.16 \text{ m}^3/\text{s}$$
   The 10^4 changes hectares to m^2; the 100 changes cm to m; the 3600 changes hours to seconds. Find i on the given figure.

2. **C** The slope is:
$$S = \Delta z / \Delta L = (201 - 200) / 180 = 0.005556$$

3. **D** Use Manning's equation on p. 40 to find the diameter assuming the pipe flows full:
$$V = Q / \pi D^2 / 4 = (1/n)(D/4)^{2/3} S^{1/2}$$
$$4.16 / \pi D^2 / 4 = (1/0.015)(D^{2/3}/4^{2/3}) 0.00556^{1/2} \quad \therefore D = 1.45 \text{ m}$$
   We used the hydraulic radius $R = D/4$.

Example 7.2

A 2-m-diameter, 200-m-long, cast-iron pipe transports water from a reservoir with surface elevation 726 m to an 89%-efficient turbine which has its outlet at 696 m. The turbine operates such that the flow rate is 6 m^3/s. Use kinematic viscosity $\nu = 10^{-6}$ m^2/s.

1. A 15-m-wide, 1.2-m deep river feeds the reservoir from above. Estimate the river's slope if the Manning n is 0.035.
   (A) 0.013
   (B) 0.0013
   (C) 0.00013
   (D) 0.000013

2. Approximate the losses up to the inlet of the turbine.
   (A) 20 m
   (B) 10 m
   (C) 2.5 m
   (D) 0.25 m

3. What is the expected output of the turbine?
   (A) 1560 kW
   (B) 960 kW
   (C) 270 kW
   (D) 43.2 kW

4. Estimate the maximum possible velocity entering a blade on the turbine.
   (A) 15 m/s
   (B) 28 m/s
   (C) 37 m/s
   (D) 43 m/s

Solutions:

1. **C** Use the Manning equation on p. 40 of the Handbook:
$$Q = \frac{1}{n} R^{2/3} S^{1/2} A \quad \text{or} \quad 6 = \frac{1}{0.035}\left(\frac{15 \times 1.2}{2.4 + 15}\right)^{2/3} \sqrt{S}. \quad \therefore S = 0.00013$$

2. **D** To find the head loss we must know the velocity, the Reynolds number and the e/D ratio:
$$V = Q/A = 6/\pi \times 1^2 = 1.91 \text{ m/s}. \quad e/D = 0.25/2000 = 0.000125.$$
$$\text{Re} = \frac{VD}{\nu} = \frac{1.91 \times 2}{10^{-6}} = 3.8 \times 10^{-6}.$$
$\therefore$ from Moody diagram: $f = 0.013$

Finally,
$$h_f = f\frac{L}{D}\frac{V^2}{2g} = 0.013 \frac{200}{2} \frac{1.91^2}{2 \times 9.8} = 0.24 \text{ m}$$

3. **A** Use the energy equation:
$$\frac{p_1}{\gamma} + z_1 + \frac{V_1^2}{2g} = \frac{p_2}{\gamma} + z_2 + \frac{V_2^2}{2g} + h_f + H_T$$
$$0 + 726 + 0 = 0 + 696 + 0 + 0.013 \frac{200}{2} \frac{1.91^2}{2 \times 9.8} + H_T. \quad \therefore H_T = 30 \text{ m}$$
$$\dot{W} = Q\gamma H_T \eta = 6 \times 9800 \times 30 \times 0.89 = 1560000 \text{ W} \quad \text{or} \quad 1560 \text{ kW}$$

We have used an expression similar to the pump power equation on p. 39, except we must multiply when efficiency is used for a turbine.

4. **B** The maximum velocity is limited by the absolute pressure going to zero (actually, it cannot be lower than the vapor pressure, but the vapor pressure is nearly absolute zero for river temperatures). The energy equation from the surface to the blade inlet provides us with the maximum velocity:
$$\frac{p_1}{\gamma} + z_1 + \frac{V_1^2}{2g} = \frac{p_2}{\gamma} + z_2 + \frac{V_2^2}{2g} + h_f$$
$$0 + 726 + 0 = \frac{-100\,000}{9800} + 696 + \frac{V_2^2}{2 \times 9.8} + 0.013 \frac{200}{2} \frac{1.91^2}{2 \times 9.8}$$
$$\therefore V_2 = 28.1 \text{ m/s}$$

Civil Engineering Discipline Exam

This practice exam has been developed to help test-takers prepare for the afternoon Discipline exam in Civil Engineering. In addition, if test-takers are undecided, taking this test may also help them determine whether they should take the afternoon DS test or the afternoon General test.

The subjects tested are those subjects that the NCEES says will make up the exam. We have placed the same number of problems from each area of Civil Engineering that can be expected to be on the actual exam. The difficulty level can, of course, be quite different on the actual exam; we have, however, attempted to create an exam with essentially the same level of difficulty as one would expect on the actual exam.

If you find the problems on this exam to be more difficult than the problems on the afternoon General test, you may wish to take the General test in the afternoon. About one-half of all Civil Engineering examinees take the afternoon General test. This practice exam should help you make that decision before the day of the test.

This exam is also included on the *free* CD which is available via the coupon at the back of this book. The Study-Director™ feature of the CD will provide an analysis of your results and help you determine which subjects to review again, should you need extra review. Do not look at this exam if you plan to take the exam on the CD.

FUNDAMENTALS OF ENGINEERING EXAM

Afternoon Session—Civil Exam

(Simulated answer form with topical breakout and scoring grid.)

BE SURE EACH MARK IS DARK AND COMPLETELY FILLS THE INTENDED SPACE AS ILLUSTRATED HERE: ●.

CONCRETE	SURVEYING	WATER TREATMENT	SOILS	CONTRACTS
1 Ⓐ Ⓑ Ⓒ Ⓓ	13 Ⓐ Ⓑ Ⓒ Ⓓ	25 Ⓐ Ⓑ Ⓒ Ⓓ	37 Ⓐ Ⓑ Ⓒ Ⓓ	49 Ⓐ Ⓑ Ⓒ Ⓓ
2 Ⓐ Ⓑ Ⓒ Ⓓ	14 Ⓐ Ⓑ Ⓒ Ⓓ	26 Ⓐ Ⓑ Ⓒ Ⓓ	38 Ⓐ Ⓑ Ⓒ Ⓓ	50 Ⓐ Ⓑ Ⓒ Ⓓ
3 Ⓐ Ⓑ Ⓒ Ⓓ	15 Ⓐ Ⓑ Ⓒ Ⓓ	27 Ⓐ Ⓑ Ⓒ Ⓓ	39 Ⓐ Ⓑ Ⓒ Ⓓ	51 Ⓐ Ⓑ Ⓒ Ⓓ
4 Ⓐ Ⓑ Ⓒ Ⓓ	16 Ⓐ Ⓑ Ⓒ Ⓓ	28 Ⓐ Ⓑ Ⓒ Ⓓ	40 Ⓐ Ⓑ Ⓒ Ⓓ	52 Ⓐ Ⓑ Ⓒ Ⓓ
5 Ⓐ Ⓑ Ⓒ Ⓓ	17 Ⓐ Ⓑ Ⓒ Ⓓ		41 Ⓐ Ⓑ Ⓒ Ⓓ	53 Ⓐ Ⓑ Ⓒ Ⓓ
6 Ⓐ Ⓑ Ⓒ Ⓓ	18 Ⓐ Ⓑ Ⓒ Ⓓ		42 Ⓐ Ⓑ Ⓒ Ⓓ	54 Ⓐ Ⓑ Ⓒ Ⓓ
Score: _____	Score: _____	Score: _____	Score: _____	Score: _____

STEEL	HYDRAULICS/ HYDROLOGY	ENVIRONMENTAL	TRANSPORTATION	COMPUTERS
7 Ⓐ Ⓑ Ⓒ Ⓓ	19 Ⓐ Ⓑ Ⓒ Ⓓ	29 Ⓐ Ⓑ Ⓒ Ⓓ	43 Ⓐ Ⓑ Ⓒ Ⓓ	55 Ⓐ Ⓑ Ⓒ Ⓓ
8 Ⓐ Ⓑ Ⓒ Ⓓ	20 Ⓐ Ⓑ Ⓒ Ⓓ	30 Ⓐ Ⓑ Ⓒ Ⓓ	44 Ⓐ Ⓑ Ⓒ Ⓓ	56 Ⓐ Ⓑ Ⓒ Ⓓ
9 Ⓐ Ⓑ Ⓒ Ⓓ	21 Ⓐ Ⓑ Ⓒ Ⓓ	31 Ⓐ Ⓑ Ⓒ Ⓓ	45 Ⓐ Ⓑ Ⓒ Ⓓ	57 Ⓐ Ⓑ Ⓒ Ⓓ
10 Ⓐ Ⓑ Ⓒ Ⓓ	22 Ⓐ Ⓑ Ⓒ Ⓓ	32 Ⓐ Ⓑ Ⓒ Ⓓ	46 Ⓐ Ⓑ Ⓒ Ⓓ	58 Ⓐ Ⓑ Ⓒ Ⓓ
11 Ⓐ Ⓑ Ⓒ Ⓓ	23 Ⓐ Ⓑ Ⓒ Ⓓ	33 Ⓐ Ⓑ Ⓒ Ⓓ	47 Ⓐ Ⓑ Ⓒ Ⓓ	59 Ⓐ Ⓑ Ⓒ Ⓓ
12 Ⓐ Ⓑ Ⓒ Ⓓ	24 Ⓐ Ⓑ Ⓒ Ⓓ	34 Ⓐ Ⓑ Ⓒ Ⓓ	48 Ⓐ Ⓑ Ⓒ Ⓓ	60 Ⓐ Ⓑ Ⓒ Ⓓ
		35 Ⓐ Ⓑ Ⓒ Ⓓ		
		36 Ⓐ Ⓑ Ⓒ Ⓓ		
Score: _____	Score: _____	Score: _____	Score: _____	Score: _____

Civil Discipline Exam

Questions 1-3 relate to the reinforced concrete beam shown below.

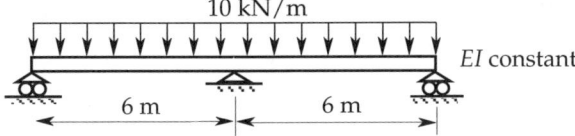

Questions 1 – 3

1. The beam is
   (A) determinate
   (B) indeterminate
   (C) improperly restrained
   (D) none of the above

2. The value of the reaction force at the center support is most nearly:
   (A) 20 kN
   (B) 25 KN
   (C) 30 kN
   (D) 75 kN

3. Tension reinforcement at the center support should be located:
   (A) near the top surface of the beam
   (B) at mid-depth
   (C) near the bottom surface of the beam
   (D) no tension reinforcement needed

Questions 4–6 relate to the cross-section of a short reinforced concrete column shown below. The 28-day compression strength is 4000 psi. Reinforcement is Grade 60. Longitudinal bars are #9, and spiral reinforcing is #3 (not shown).

Questions 4 – 6

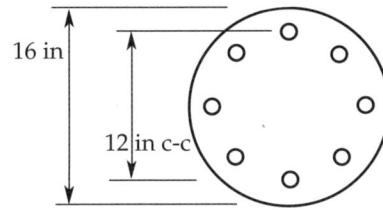

4. The reinforcement ratio is most nearly:
   (A) 0.02
   (B) 0.04
   (C) 0.06
   (D) 0.08

5. Clear cover is most nearly:
   (A) 1 inch
   (B) 1-1/2 inch
   (C) 2 inches
   (D) 3-1/2 inches

6. The load capacity of the column is most nearly:
   (A) 560 kips
   (B) 680 kips
   (C) 800 kips
   (D) 970 kips

Questions 7 - 9

Questions 7–9 relate to a flat bar tension member made of steel with tensile yield strength of 50 ksi and ultimate strength of 65 ksi. The figure below illustrates the holes for a bolted connection at the end of the member. The effective hole diameter is 1.00 inch.

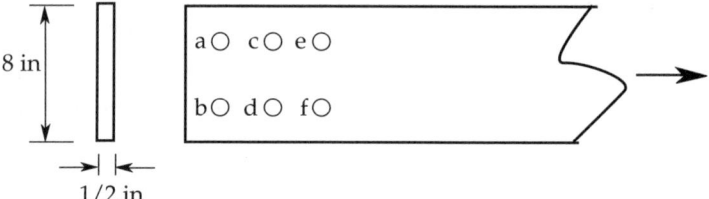

7. The net area is most nearly:
   (A) 1.0 in²
   (B) 2.5 in²
   (C) 3.0 in²
   (D) 4.0 in²

8. A block shear failure involves fracturing along a path through holes:
   (A) e-f only
   (B) a-c-e only
   (C) a-c-e-f-d-b
   (D) a-b only

9. Assume that tension yielding is the critical limit state and that dead load is 30 kips. The maximum permissible live load is most nearly:
   (A) 90 kips
   (B) 100 kips
   (C) 120 kips
   (D) 150 kips

Questions 10-12 relate to the cross section of a built-up compression member shown below:

Questions 10 – 12

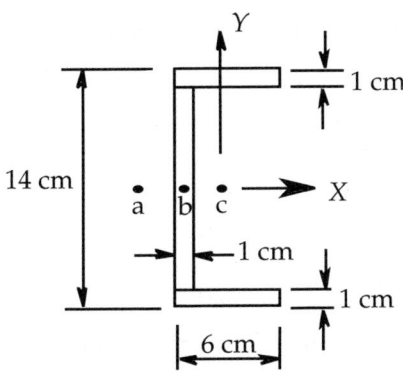

10. The shear center is most nearly located at point:
    (A) a
    (B) b
    (C) c
    (D) This section does not have a shear center

11. The moment of inertia about the X-axis is most nearly:
    (A) 5 cm^4
    (B) 24 cm^4
    (C) 200 cm^4
    (D) 650 cm^4

12. Assume that the moment of inertia about the Y-axis is 74.5 cm^4, that the member is 2.40 meters long, and that the ends are restrained from both rotation and lateral translation. The slenderness ratio for buckling about the Y-axis is most nearly:
    (A) 20
    (B) 90
    (C) 200
    (D) 430

Questions 13 – 18 Questions 13-18 refer to the plat of a suburban lot shown below. The front boundary is a circular curve and the side boundaries are radial to that curve. The table gives plane coordinates of the corners of the lot. Units can be feet or meters.

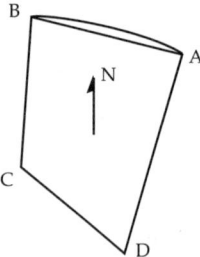

Pt	N	E
A	339.74	310.09
B	391.50	116.90
C	198.29	100.00
D	100.00	198.30

13. The angle of intersection of the curve is most nearly:
    (A) 10°
    (B) 20°
    (C) 70°
    (D) 80°

14. The chord length AB is most nearly
    (A) 193
    (B) 200
    (C) 201
    (D) 207

15. The radius of the curve is most nearly:
    (A) 400
    (B) 567
    (C) 576
    (D) 585

16. The curve length AB is most nearly:
    (A) 193
    (B) 200
    (C) 201
    (D) 207

17. The bearing of line AB is most nearly:
    (A) N 15° W
    (B) W 15° N
    (C) N 60° W
    (D) N 75° W

18. The area bounded by traverse ABCD is most nearly:
    (A) 36,400
    (B) 43,560
    (C) 50,000
    (D) 72,800

A construction project requires that water be delivered to a small reservoir at the site at elevation 400 meters. Water is available in a lake at elevation 390 meters. There is 4-cm-diameter, galvanized iron piping and a pump available. The pump has characteristics shown. It will require 400 meters of pipe. Use $v_{water} = 10^{-6}$ m²/s.

Questions 19 – 22

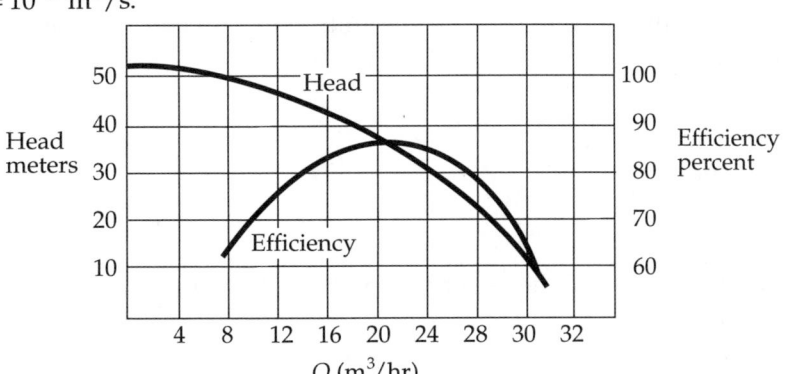

19. In the analysis of the piping system, we:
    (A) must include minor losses
    (B) may neglect minor losses
    (C) must include the pressure drop from inlet to exit
    (D) must include the kinetic energy at the exit

20. If the flow is controlled with a valve to be 6 m³/hr, the head loss over 400 meters of pipe is most nearly:
    (A) 27 m
    (B) 37 m
    (C) 47 m
    (D) 57 m

21. Estimate the flow rate assuming the water flows freely through the pipe from the river to the reservoir:
    (A) 8 m³/hr
    (B) 12 m³/hr
    (C) 16 m³/hr
    (D) 20 m³/hr

22. The pump can be placed anywhere along the length of the pipe. If it is placed just before the reservoir it may not perform because:
    (A) pump efficiency is too low
    (B) turbulence is too high
    (C) surface tension prevents proper flow
    (D) vapor pressure may be a factor

Questions 23–24

The runoff coefficient for a particular 50 acre industrial area is determined to be 0.8. A storm produces rain intensity of 0.5 in./hr. A 3-ft-diameter storm sewer transports the water from the area. The Manning roughness coefficient is 0.012.

23. Estimate the velocity of the water in the storm sewer if it flows full with no surcharge.
    (A) 2 fps
    (B) 3 fps
    (C) 4 fps
    (D) 5 fps

24. What slope should the sewer have if it is to flow full with no surcharge?
    (A) 0.002
    (B) 0.0012
    (C) 0.00029
    (D) 0.00077

Questions 25 – 28

A partial ion analysis has been performed on a groundwater that is to be treated using conventional lime/ soda ash softening. The flow rate of the water to be treated is 15 MGD. The ion concentrations reported below are all given in units of mg/L as $CaCO_3$. The pH of the water is 7.8.

$$Ca^{2+} = 240.0$$
$$Mg^{2+} = 60.0$$
$$Fe^{3+} = 1.3$$
$$Mn^{4+} = 0.65$$
$$Na^+ = 45.0$$
$$HCO_3^- = 220.0$$
$$SO_4^{2-} = 90.0$$
$$CO_2 = 15.0$$

25. Calculate the total hardness of this water.
    (A) 302 mg/L as $CaCO_3$
    (B) 347 mg/L as $CaCO_3$
    (C) 180 mg/L as $CaCO_3$
    (D) 547 mg/L as $CaCO_3$

26. Calculate the carbonate hardness.
    (A) 122 mg/L as $CaCO_3$
    (B) 167 mg/L as $CaCO_3$
    (C) 180 mg/L as $CaCO_3$
    (D) 220 mg/L as $CaCO_3$

27. Calculate the amount of lime (in kg) that must be added each day to react with the carbon dioxide present in the water.
    (A) 12 618 kg/d
    (B) 31 544 kg/d
    (C) 631 kg/d
    (D) 7565 kg/d

28. Calculate the amount of soda ash that must be required to soften this water.
    (A) 3610 kg/d
    (B) 7220 kg/d
    (C) 903 kg/d
    (D) 4815 kg/d

Questions 29 – 32

A municipal wastewater treatment plant is being designed to treat wastewater at a flow rate of 0.75 m³/s. A horizontal flow grit chamber is to be used to pretreat the wastewater. The approach velocity in the chamber is to be 0.35 m/s. The length of the chamber is to be 15 times the width and the depth is to be 1.8 m.

A circular primary sedimentation basin is to be designed for this facility. The tank is designed to have an overflow rate of 40 m³/m²/day and a detention time of 2.0 h. The weir loading rate is to be 250 m³/m/day.

29. Calculate the width of the grit chamber.
    (A) 0.13 m
    (B) 0.95 m
    (C) 1.2 m
    (D) 2.4 m

30. Calculate the volume of the grit chamber.
    (A) 0.46 m³
    (B) 24.4 m³
    (C) 78 m³
    (D) 39 m³

31. Calculate the diameter of the sedimentation basin.
    (A) 23 m
    (B) 18 m
    (C) 9.1 m
    (D) 45 m

32. Calculate the length of weir required.
    (A) 1620 m
    (B) 810 m
    (C) 260 m
    (D) 130 m

Questions 33 – 36 The Westwood Wastewater Treatment Plant (WWTP) uses primary sedimentation followed by activated sludge. Ferric chloride is added at the end of the activated sludge basin to precipitate phosphate. Following secondary clarification, tertiary filtration is used to remove turbidity. The final effluent is treated with chlorine for disinfection followed by dechlorination using sodium bisulfite. Just before discharge into the Red Oak River the wastewater is aerated to increase the dissolved oxygen level.

33. The main sewer entering the Westwood WWTP is 4 ft in diameter. The sewer flows at full capacity at a flow rate of 7.5 ft^3/s and a velocity of 0.6 ft/s. What is the depth of water in the sewer when the wastewater flow is at a flow rate of 5.25 ft^3/s? (Note: Assume that the Manning's constant, n, is constant with depth.)

    (A) 34 in
    (B) 32 in
    (C) 29 in
    (D) 24 in

34. The influent wastewater has BOD$_5$ and suspended solids (SS) concentrations of 240 mg/L and 260 mg/L, respectively. Primary sedimentation results in a 70% decrease in the suspended solids concentration and a 30% decrease in the BOD$_5$. Calculate the amount of primary sludge produced each day, assuming a flow rate of 7.5 ft^3/s.

    (A) 4700 kg/d
    (B) 3300 kg/d
    (C) 10,900 kg/d
    (D) 150 kg/d

35. What should be the volume of the aeration basin (for activated sludge treatment) if the organic loading rate (volumetric) is to be 75 lb BOD$_5$/1000 ft^3-d. Use the peak flow rate of 7.5 ft^3/s.

    (A) 12,000 ft^3
    (B) 4330 ft^3
    (C) 28,000 ft^3
    (D) 90,500 ft^3

36. The water flow meter at the Westwood water plant is not operating. The plant superintendent tells you that each of the four dual media filters (each 10.00 ft × 20.0 ft) are loaded at a velocity of 600 ft/d. The depth of the filters is 9.00 ft. What is the total flow rate through the four filters?

    (A) 5.6 ft^3/s
    (B) 1.4 ft^3/s
    (C) 5.0 ft^3/s
    (D) 2.5 ft^3/s

Questions 37 – 39

A saturated, normally consolidated clay has a water content of 30 percent, a liquid limit of 40 percent, and a plastic limit of 20 percent. The specific gravity of solids is 2.70.

37. Estimate the compression index for this soil.
    (A) 0.300
    (B) 0.360
    (C) 0.270
    (D) 0.0027

38. Estimate the void ratio for this soil.
    (A) 1.080
    (B) 0.810
    (C) 0.300
    (D) 0.429

39. The vertical stress increase caused by a newly-constructed building causes the effective stress on this soil in the ground to double. How much will the void ratio of this clay decrease as the building settles?
    (A) 0.187
    (B) 0.081
    (C) 0.013
    (D) 0.500

Questions 40 – 42

A retaining wall is as shown. The sand in the backfill and foundation has a unit weight of 125 lb/ft³ and a friction angle of 32 degrees.

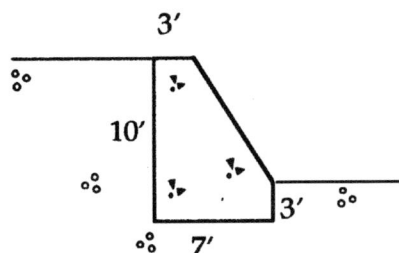

40. What is the total active earth force per one-foot length of wall?
    (A) 417 lb
    (B) 1919 lb
    (C) 1250 lb
    (D) 384 lb

41. What is the overturning moment per one-foot length of wall?
    (A) 6396 ft-lb
    (B) 19,190 ft-lb
    (C) 9594 ft-lb
    (D) 1279 ft-lb

42. If the wall is made of concrete with a unit weight of 150 lb/ft³, what is the factor of safety against sliding? Take the factor of safety as the ratio of the resisting forces to the driving forces.
   (A) 2.00
   (B) 1.05
   (C) 0.95
   (D) 3.69

Questions 43 – 44 A 2-directional 24-foot wide asphalt concrete (AC) surfaced road is to be constructed along with a total of 12-foot asphalt concrete shoulders. The road and shoulders cross section includes 12-inch granular subbase, 6-inch gravel base, and 6-inch asphalt concrete. The asphalt paver places the AC in two courses: 4-inch base course and 2-inch surface course. The paver places the AC in 12-ft wide strips at an average speed of 2 miles per day.

43. If the density of the placed and compacted AC is 145 pcf, then the total weight in pounds of the AC material (both surface and base courses) in a one-square-yard surface area is about:
   (A) 752.5
   (B) 652.5
   (C) 600.0
   (D) 700.0

44. The density of the AC material in the trucks is 120 pcf. The daily volume (the number of cubic yards) of the AC surface course in the delivery trucks is about:
   (A) 1045
   (B) 1245
   (C) 945
   (D) 2145

45. The new road was designed and constructed to carry the 20-year design period of traffic. The projected two-directional traffic in terms of 18,000 pounds equivalent single axle load (ESAL) over the 20-year design period was 2,900,000 ESAL. The design ESAL (the number of ESAL used in the design of the road) is about:
   (A) 2,900,000
   (B) 1,450,000
   (C) 5,800,000
   (D) None of the above

A freeway in a rural area has the following design criteria:

Questions 46 – 49

Design Speed	70 mph
Wet Coefficient of Friction	0.28
Perception of Reaction Time	2.5 secs
Height of Eye	3.5 ft
Height of Object	0.5 ft
Top of Vehicle	4.25 ft

46. The distance traveled by a vehicle during the perception and reaction of the driver before stopping is most nearly:
    (A) 260 ft
    (B) 290 ft
    (C) 320 ft
    (D) 350 ft

47. The distance traveled by the vehicle while decelerating to a stop (leaving skid marks) is most nearly:
    (A) 500 ft
    (B) 550 ft
    (C) 580 ft
    (D) 620 ft

48. If the maximum allowable superelevation and lateral friction force are 0.08 and 0.10, respectively, the minimum design radius is most nearly:
    (A) 1745 ft
    (B) 1815 ft
    (C) 1910 ft
    (D) 1975 ft

49. During the construction of a concrete framed office building, it is discovered that a concrete beam has insufficient steel reinforcement to support the code required floor loads. After review, it is determined that the contractor has installed the reinforcement as detailed in the contract documents and that the consulting engineer inadvertently had placed the wrong reinforcing information on the contract drawings. A general note on the drawings indicated that "the contractor shall install all work in accordance with all codes and ordinances." The contractor claims that he deserves extra compensation to replace the under-reinforced beam. The engineer argues that the "compliance with laws" note places the responsibility to install the work to meet code on the contractor and therefore the contractor is required to comply regardless of the information presented on the contract drawings. The engineer rules that the contractor must replace the beam without additional compensation. Which statement below best reflects the proper allocation of liability in this case?
    (A) The contractor shall remove and replace the beam without extra compensation because of the "compliance with laws" note.
    (B) The contractor shall remove and replace the beam and shall be compensated by the owner for the extra work.
    (C) The contractor shall remove and replace the beam and shall be compensated by the owner for the extra work. The owner may then attempt to obtain reimbursement from the engineer for the extra work required by the engineer's incorrect drawing.
    (D) The engineer and contractor are equally responsible and should share in the cost of removing and replacing the beam.

50. During negotiations between an owner and consulting engineer for the provision of engineering design services, the owner insists on including the following language in the contract:

    "The engineer shall provide engineering design services in a timely manner and with the highest standard of care of his or her profession."

    The engineer should do which of the following?
    (A) Recognize that owners have a right to receive quality services and agree to the clause without modification.
    (B) Refuse to enter into the agreement because of the "timely manner" provision.
    (C) Refuse to enter into the agreement because of the "standard of care" provision.
    (D) Attempt to negotiate a new clause with a lower standard of care provision.

51. A "general contract" is let for the construction of an electrical power substation. Which answer below best describes the contractual lines of privity between the parties in this form of construction project management?
    (A) The consulting engineer will have a contract with the owner, but will not have a contract with either the general contractor or the subcontractors.
    (B) The consulting engineer will have a contract with the owner, and also a separate contract with the general contractor.
    (C) The consulting engineer will have separate contracts with the owner, general contractor, and the subcontractors.
    (D) The consulting engineer will have a contract with the general contractor but will not have agreements with either the owner or the subcontractors.

52. Which statement is true regarding the legality, enforceability and binding of contracts?
    (A) For a contract to be enforceable it must always be in writing.
    (B) Contracts written for an illegal purpose will still be enforced by a court.
    (C) Consideration must be evident in the contract. Consideration refers to fairness of the agreement.
    (D) Mutual agreement of the parties must be evident.

53. Often in legal disputes involving construction and engineering projects the courts will use past precedent setting legal cases to help interpret and decide the case before the court. This general body of law, which has been developed through decisions of courts, is referred to as which of the following?
    (A) Tort Law
    (B) Statute Law
    (C) Common Law
    (D) Public Law

54. Which of the duties below would not be considered as a construction contract administration duty of a consulting engineer?
    (A) Preparation of initial design drawings
    (B) Field observation of the construction project
    (C) Review of the contractor's payment requests
    (D) Processing of contract change orders

55. A spreadsheet has cell $B2$ set to 4. Then cell $B3$ is set to the formula $\$B\$2 + B2$. This formula is copied into cells $B4, B5, B6$, and $B7$ in order. The number in $B6$ is:
    (A) 20
    (B) 24
    (C) 32
    (D) 16

56. The pseudo-code segment is given the following list of integer values as data: 17, 15, 12, 18, 19, 25. After execution, what values will be displayed?

$$a := 100;\ b := 0;\ c := 0;$$
$$\text{for } i := 1 \text{ to } 5$$
$$\quad \text{read}(x);$$
$$\quad \text{if}(x < a)$$
$$\quad\quad a := x;$$
$$\quad \text{if}(x > b)$$
$$\quad\quad b := x;$$
$$\quad c := c + x;$$
$$\text{endfor}$$
$$\text{print}(a, b, c/5);$$

(A) 12, 25, 21
(B) 12, 17, 16
(C) 12, 19, 16
(D) 12, 19, 21

57. A file contains the integers: 3, 2, –4, 5, 1, –2, ... which are processed by the pseudo-code program:

$$a := 0;$$
$$i := 0;$$
$$\text{while}(i \leq 3)$$
$$\quad \text{read}(b);$$
$$\quad \text{if}(b > 0)$$
$$\quad\quad a = a + b*b;$$
$$\quad i := i + 1;$$
$$\text{endwhile};$$
$$\text{print}(a/i);$$

What will be the approximate value printed?
(A) 11
(B) 11.2
(C) 7.8
(D) 9.7

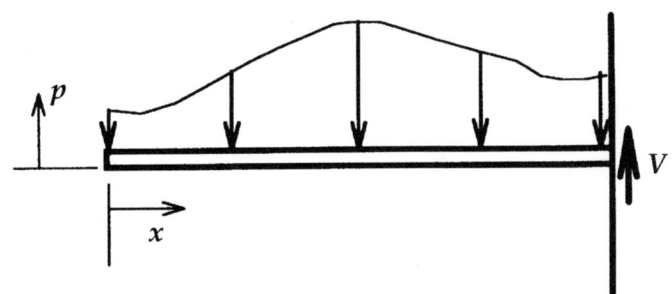

Questions 58 – 59

A 48-inch-long beam is loaded with a varying distributed load p which is a function of distance x along the beam. The shear V is the integral of p with respect to x, or the area under the $p(x)$ curve. Values of p at points along the beam are as follows:

x, inches	0	12	24	36	48
p, lb/in	5	11	18	14	10

58. Determine V using the forward trapezoidal rule.
    (A) 606 lb
    (B) 604 lb
    (C) 576 lb
    (D) 557 lb

59. Determine V using Simpson's rule.
    (A) 606 lb
    (B) 604 lb
    (C) 576 lb
    (D) 557 lb

60. A root of $P(x) = x^3 - 4x + 4$ near $x = -1$ is most nearly:
    (A) −0.949
    (B) −1.043
    (C) −1.099
    (D) −0.903

FUNDAMENTALS OF ENGINEERING EXAM

Afternoon Session—Civil Exam

(Simulated answer form with topical breakout and scoring grid.)

BE SURE EACH MARK IS DARK AND COMPLETELY FILLS THE INTENDED SPACE AS ILLUSTRATED HERE: ●.

CONCRETE	SURVEYING	WATER TREATMENT	SOILS	CONTRACTS
1 Ⓐ●ⒸⒹ	13 Ⓐ●ⒸⒹ	25 ●ⒷⒸⒹ	37 ⒶⒷ●Ⓓ	49 ⒶⒷ●Ⓓ
2 ⒶⒷⒸ●	14 Ⓐ●ⒸⒹ	26 ⒶⒷ●Ⓓ	38 Ⓐ●ⒸⒹ	50 ⒶⒷⒸ●
3 ●ⒷⒸⒹ	15 ⒶⒷ●Ⓓ	27 ⒶⒷ●Ⓓ	39 Ⓐ●ⒸⒹ	51 ●ⒷⒸⒹ
4 Ⓐ●ⒸⒹ	16 ⒶⒷ●Ⓓ	28 ⒶⒷⒸ●	40 Ⓐ●ⒸⒹ	52 ⒶⒷⒸ●
5 ●ⒷⒸⒹ	17 ⒶⒷⒸ●		41 ●ⒷⒸⒹ	53 ⒶⒷ●Ⓓ
6 ⒶⒷⒸ●	18 Ⓐ●ⒸⒹ		42 ⒶⒷⒸ●	54 ●ⒷⒸⒹ
Score: _____	Score: _____	Score: _____	Score: _____	Score: _____

STEEL	HYDRAULICS/ HYDROLOGY	ENVIRONMENTAL	TRANSPORTATION	COMPUTERS
7 ⒶⒷ●Ⓓ	19 Ⓐ●ⒸⒹ	29 ⒶⒷ●Ⓓ	43 Ⓐ●ⒸⒹ	55 ●ⒷⒸⒹ
8 ⒶⒷ●Ⓓ	20 ●ⒷⒸⒹ	30 ⒶⒷⒸ●	44 ⒶⒷ●Ⓓ	56 ⒶⒷ●Ⓓ
9 ●ⒷⒸⒹ	21 ●ⒷⒸⒹ	31 ⒶⒷⒸ●	45 Ⓐ●ⒸⒹ	57 ⒶⒷⒸ●
10 ●ⒷⒸⒹ	22 ⒶⒷⒸ●	32 ⒶⒷ●Ⓓ	46 ●ⒷⒸⒹ	58 ●ⒷⒸⒹ
11 ⒶⒷⒸ●	23 Ⓐ●ⒸⒹ	33 ●ⒷⒸⒹ	47 ⒶⒷ●Ⓓ	59 Ⓐ●ⒸⒹ
12 Ⓐ●ⒸⒹ	24 ⒶⒷ●Ⓓ	34 Ⓐ●ⒸⒹ	48 Ⓐ●ⒸⒹ	60 ⒶⒷⒸ●
		35 ⒶⒷⒸ●		
		36 ●ⒷⒸⒹ		
Score: _____	Score: _____	Score: _____	Score: _____	Score: _____

Solutions to the Civil Exam

1. **B** Number of reactions > number of independent equations of static equilibrium.

2. **D** By flexibility method:
$$(5wL^4)/(384EI) = (PL^3)/(48EI)$$

   Therefore,
   $$P = 0.625wL$$
   $$= 0.625 \times 10 \text{ kN/m} \times 12 \text{ m} = 75 \text{ kN}$$

3. **A** Moment is negative at center, so the top of the beam is in tension.

4. **B**
$$A_g = \pi d^2/4 = 201 \text{ in}^2$$
$$A_s = 8 \times 1.00 = 8.0 \text{ in}^2$$
$$\rho_g = A_s/A_g = 8.00/201 = 0.04$$

5. **A** $16/2 - 12/2 - 1.128/2 - 0.375 = 1.06$ inch

6. **D**
$$A_{concrete} = A_g - A_s = 201 - 8 = 193 \text{ in}^2$$
$$P_o = 0.85 f_c' A_{concrete} + f_y A_s$$
$$= 0.85 \times 4 \text{ ksi} \times 193 \text{ in}^2 + 60 \text{ ksi} \times 8 \text{ in}^2 = 1136 \text{ kips}$$
$$P_n = \phi P_o = 0.85 \times 1136 = 966 \text{ kips}$$

7. **C**
$$A_n = b_n t = (b - \Sigma d)t$$
$$= (8.0 - 2 \times 1.0) \times 0.5 = 3.0 \text{ in}^2$$

8. **C** By definition of block shear.

9. **A** By ASD:
$$F_t = 0.6 F_y = 0.6 \times 50 = 30 \text{ ksi}$$
$$P_n = F_t A_g = 30 \text{ ksi} \times 4.0 \text{ in}^2 = 120 \text{ kips}. \quad \therefore \ 120 - 30 = 90 \text{ kips}$$

   By LRFD:
$$\phi P_n = 0.9 F_y A_g = 0.9 \times 50 \text{ ksi} \times 4.0 \text{ in}^2 = 180 \text{ kips}$$
$$1.2D + 1.6L \le \phi P_n$$

   Therefore,
$$L \le (\phi P_n - 1.2D)/(1.6 = 180 - 1.2 \times 30)/1.6 = 90 \text{ kips}$$

10. **A**

11. **D** $(6 \times 14^3)/12 - (5 \times 12^3)/12 = 652 \text{ cm}^4$

12. **B** $$r = \sqrt{I/A} = \sqrt{74.5/24} = 1.76 \text{ cm}$$
    $$KL/r = 0.65 \times 240/1.76 = 89$$

13. **B** $AzBC = \tan^{-1}[(X_C - X_B)/(Y_C - Y_B)] =$
    $\tan^{-1}[(100.00 - 116.90)/(198.29 - 391.50)] = 185°$
    $AzAD = \tan^{-1}[(198.30 - 310.09)/(100.00 - 339.74)] = 205°$ ∴ $205 - 185 = 20°$

14. **B**
    $$\sqrt{(X_B - X_A)^2 + (Y_B - Y_A)^2}$$
    $$= \sqrt{(116.90 - 310.09)^2 + (391.50 - 339.74)^2} = 200$$

15. **C** $(L.C.)/2\sin(I/2) = 200.00/2\sin 10° = 575.88$

16. **C** $L = RI = 575.88 \times 20\pi/180 = 201.02$

17. **D** $\tan^{-1}[(116.90 - 310.09)/(391.50 - 339.74)] = -75°$. ∴ N 75° W

18. **A** $[X_A(Y_B - Y_D) + X_B(Y_C - Y_A) + ...]/2 =$
    $[310.09(391.50 - 100.00) + 116.90(198.29 - 339.74) + 100.00(100.00 - 391.50)$
    $+ 198.30(339.74 - 198.29)]/2 = 36\,378$

19. **B** The relative roughness is $e/D = 0.15/40 = 0.00375$. Assuming $Re > 10^5$, or so, $f \cong 0.026$. If fL/D is quite large we can neglect the minor losses:
    $$f\frac{L}{D} = 0.026\frac{400}{0.04} = 260$$
    Therefore, minor losses are negligible; we know that $C_{minor} \leq 1$ (see p.37 of the Handbook). Compare C_{minor} with fL/D.

20. **A** Calculate the friction factor using the Moody diagram:
    $$V = \frac{Q}{A} = \frac{6/3600}{\pi \times 0.02^2} = 1.33 \text{ m/s}, \quad Re = \frac{VD}{v} = \frac{1.33 \times 0.04}{10^{-6}} = 5.3 \times 10^4.$$
    $$\frac{e}{D} = 0.00375. \quad \therefore f = 0.03.$$
    Then, the head loss is
    $$h_f = f\frac{L}{D}\frac{V^2}{2g} = 0.03\frac{400}{0.04}\frac{1.33^2}{2 \times 9.8} = 27 \text{ m}.$$

21. **A** Apply the energy equation (we use Q as m³/hr):

$$\frac{p_1}{\gamma} + z_1 + \frac{V_1^2}{2g} + h_P = \frac{p_2}{\gamma} + z_2 + \frac{V_2^2}{2g} + h_f$$

$$0 + 390 + 0 + h_P = 0 + 400 + 0 + 0.028 \frac{400}{0.04} \frac{Q^2/(\pi \times 0.02^2)}{2 \times 9.8 \times 3600^2}$$

$$h_P = 10 + 0.7Q^2$$

Solve this system equation and the characteristic curve simultaneously by trial-and-error:

try $h_P = 40$: $Q = 6.56$ (from equation)

$Q = 19$ (from curve)

try $h_P = 50$: $Q = 7.56$ (from equation)

$Q = 8$ (from curve)

The approximate flow rate is $Q = 8$ m^3/hr.

22. **D** The pressure decreases along the pipe up to the pipe inlet. Cavitation (vaporization of the water) might result if the inlet pressure falls below the vapor pressure. This condition must be checked to avoid possible cavitation.

23. **B** The average velocity of water in the sewer is

$$V = \frac{Q}{A} = \frac{ciA_{surface}}{A_{pipe}} = \frac{0.8 \times 0.5 \times 50}{\pi \times 1.5^2} = 2.83 \text{ fps}$$

Note: The coefficient C requires the surface area to be in acres.

24. **D** Use Manning's equation:

$$V = \frac{1.486}{n} R^{2/3} S_0^{1/2}$$

$$2.83 = \frac{1.486}{0.012} \left(\frac{\pi \times 1.5^2}{\pi \times 3} \right)^{2/3} S_0^{1/2}. \quad \therefore \quad S_0 = 0.000766$$

25. **A** The total hardness is the sum of the concentrations of the polyvalent cations:

$$[Ca^{2+}] + [Mg^{2+}] + [Fe^{3+}] + [Mn^{4+}] = 240 + 60 + 1.3 + 0.65 =$$

302 mg/L as CaCO$_3$

26. **D** The carbonate hardness is equal to the total hardness or the alkalinity, whichever is less. At a pH of 7.8, the alkalinity is approximately equal to the concentration of bicarbonate. As such, the alkalinity in this water is 220 mg/L. In this situation, since the alkalinity is less than the total hardness and the carbonate hardness cannot exceed the alkalinity, the carbonate hardness equals 220 mg/L.

27. **C** We must first calculate the concentrations of carbon dioxide in units of meq/L using the data provided in the table of equivalent weights given on p. 92 of the NCEES Handbook, 3rd ed.:

$$[CO_2] = \frac{15.0}{50.04} = 0.3 \text{ meq/L}$$

The amount of lime required can be calculated using equations 1 (Lime-Soda Softening Equations) on p. 92 of the NCEES Handbook:

	meq/L	meq/L lime required
CO_2:	0.30	0.30

Mass of lime per day =

(0.30 meq $Ca(OH)_2$/L) × (37.046 mg $Ca(OH)_2$/meq) × (3.785 L/gal) × (15 × 10^6 gal/day)(kg/10^6 mg) = 631 kg/day

28. **D** We must first calculate how much calcium and magnesium is associated with sulfate. Since the concentration of calcium is 240 mg/L as $CaCO_3$ and the alkalinity is 220 mg/L as $CaCO_3$, we can assume that the portion of the calcium that is not associated with carbonate or bicarbonate is complexed with sulfate:

$$[Ca^{2+}] = \frac{20.0}{50.04} = 0.4 \text{ meq/L}$$

The concentration of magnesium is 60 mg/L. Since the sulfate concentration is greater than the sum of the calcium that is associated with sulfate plus the magnesium, we can assume that all of the magnesium is associated with sulfate:

$$[Mg^{2+}] = \frac{60.0}{50.04} = 1.2 \text{ meq/L}$$

The amount of soda ash required is calculated from the stoichiometry of equation 3 and 5 in the Section on Lime-Soda Ash Softening Equations on p. 92 of the NCEES Handbook:

	meq/L	meq/L soda ash required
$CaSO_4$	0.40	0.40
$MgSO_4$	1.20	1.20
		1.60

Mass of soda ash per day =

(1.60 meq Na_2CO_3/L) × (53 mg Na_2CO_3/meq) × (3.785 L/gal) × (15 × 10^6 gal/day) × (kg/10^6 mg) = 4815 kg/day

29. **C** The width of the grit chamber is calculated using the equation for the approach velocity (p. 92 of the NCEES Handbook):

Approach velocity $= Q/A_x$

The approach velocity was given as 0.35 m/s. The flow rate was stipulated to be 0.75 m³/s. Therefore, the cross-sectional area A_x can be calculated by dividing the flow rate by the approach velocity:

$$A_x = \frac{Q}{V} = \frac{0.75}{0.35} = 2.14 \text{ m}^2$$

The cross-sectional area, i.e., the area of the basin that is perpendicular to the direction of flow, is also equal to the depth D times the width W:

$$A_x = D \times W = 1.8W = 2.14 \text{ m}^2$$

Therefore, $W = 1.2$ m.

30. **D** The volume of the grit chamber is

$$V = LA_x = LDW = (15 \times 1.2) \times 1.8 \times 1.2 = 39 \text{ m}^3$$

where V = volume of the basin and L = the length of the grit chamber.

31. **D** The diameter of the sedimentation is calculated from the cross-sectional area of the basin. The cross-sectional area is determined using the equation for the hydraulic loading rate given on p. 92 of the NCEES Handbook. The hydraulic loading rate or overflow rate was given to be 40 m³/m²·day.

$$A = \frac{Q}{\text{hydraulic loading rate}} = \frac{0.75 \times 3600 \times 24}{40} = 1620 \text{ m}^2$$

$$D^2 = 4A/\pi \quad \text{so that} \quad D = \sqrt{4 \times 1620/\pi} = 45 \text{ m}$$

32. **C** The length of the weir is calculated using the weir loading rate. The appropriate equation is given on p. 92 of the NCEES Handbook:

Weir loading rate $= Q/L$

where L = length of the weir:

$$L = (0.75)(3600 \times 24)/250 = 260 \text{ m}$$

33. **C** Use the chart provided on p. 98 of the NCEES Handbook. The ratio of the flows (Q/Q_f = 5.25/7.5) is 0.70. Enter this value on the chart and read off the corresponding values for the depth of flow. Use the dotted lines (for constant Darcy-Weisbach friction factor f and Manning's constant n):

Depth = 0.6 × full-flow depth = (0.6)(48 in) = 29 in

34. **B** The sludge production rate can be calculated using the flow rate, the influent SS concentration, and the percent removal:

Sludge production rate =

$$(7.5 \text{ ft}^3/\text{s})(86,400 \text{ s/d})(260 \text{ mg/L})(0.70) \times (0.028317 \text{ m}^3/\text{ft}^3)$$

$$\times (1000 \text{ L/m}^3)/(10^{-6} \text{ mg/kg}) = 3300 \text{ kg/day}$$

35. **D** The volume of the aeration basin can be calculated from the volumetric organic loading rate given on p. 92 of the NCEES Handbook:

Volume =

$$\frac{(240\,mg/L)(0.70)(7.5\,ft^3/s)(86,400\,s/d)(10^{-3}\,g/mg)(1.0\,lb/454g)(28.32\,L/ft^3)}{(0.075\,lb\,BOD/ft^3 \cdot d)}$$

$$= 90{,}500\ ft^3$$

36. **A** The flow rate through the four filters can be calculated from either the equation for the overflow rate or the equation for the approach velocity (both are given on p. 92 of the NCEES Handbook):

Overflow rate = Q/A Approach velocity = Q/A_x

The velocity given is both the overflow rate and the approach velocity. Therefore, the total flow rate through all of the four filters is

$$\frac{4(10\ ft)(20\ ft)(600\ ft/d)}{86,400\ s/d} = 5.6\ ft^3/s$$

37. **C** $C_c = 0.009(LL - 10) = 0.009(40 - 10) = 0.27$

38. **B** $e = \dfrac{wG}{S} = \dfrac{0.3 \times 2.70}{1.00} = 0.81$

39. **B** Using the relation for the compression index:
$$\Delta e = C_c \Delta \log p = C_c(\log p_2 - \log p_1) = C_c(\log 2p_1 - \log p_1)$$
$$= 0.27 \log 2 = 0.081$$

40. **B** $P_a = \dfrac{1}{2}\gamma K_a H^2 = \dfrac{1}{2} \times 125 \times 0.307 \times 10^2 = 1919\ lb/ft$

41. **A** $M_o = P_a \bar{y} = 1919(10/3) = 6396\ ft\text{-}lb/ft$

42. **D** We use the relationship $FS = (W \tan \phi + P_p)/P_a$. The various variables are
$$W = 3 \times 7 \times 150 + 0.5(3+7) \times 7 \times 150 = 8400\ lb/ft$$
$$P_p = \dfrac{1}{2}\gamma K_p H^2 = 0.5 \times 125 \times 3.255 \times 3^2 = 1831\ lb/ft$$

The factor of safety is then

$$FS = \frac{8400 \tan 32° + 1831}{1919} = 3.69$$

43. **B**

44. **C**

45. **B**

46. **A** $x = 1.47vt = 1.47 \times 70 \times 2.5 = 257\ ft$

47. **C** $$d = \frac{V^2}{30f} = \frac{70^2}{30 \times 0.28} = 583 \text{ ft}$$

48. **B** $$\textit{eff} = \frac{V^2}{15R}. \quad 0.08 + 0.10 = \frac{70^2}{15R}. \quad \therefore R = 1814 \text{ ft}$$

49. **C** The contractor is entitled to rely on the accuracy of the engineer's design. The engineer is the party licensed to perform engineering design and has the duty to design his or her work in accordance with applicable codes and ordinances. Generally, construction contracts will have clauses that will require the contractor to notify the engineer if the contractor becomes aware of a design error, but this duty does not extend to a requirement to investigate or evaluate the structural design produced by the engineer for adequacy. The "compliance with laws" clause in the scenario is a general responsibility which requires the contractor to follow laws in his or her construction operations, but this clause does not make the contractor responsible for the engineer's work. Because the contract for construction is held by the owner, the owner would have to pay for the extra charges. The owner would be likely to attempt to collect reimbursement from the engineer for work made necessary by the engineer's mistake.

50. **D** In assessing negligence an engineer is generally held by the courts and society to practice at a reasonable level equal to the average standard of care within the profession and location of the service. In other words, for the purpose of evaluating negligence the court would look to what the generally accepted engineering practice would be. Engineers are not expected to provide perfect service. By agreeing to the "highest standard of care" language required by the owner, the engineer would be agreeing to provide a service quality far in excess of what may be possible and he or she would be contractually obligated to provide such a service. Owners frequently will include "highest standard of care" clauses in their contracts, and while they occasionally will truly be seeking such a service, many times they do not really intend to impose such an extreme standard of care on the engineer. When the engineer encounters such risky clauses it is best to find out what the true intent of the owner is rather than simply rejecting the contract or agreeing to the offensive language.

Providing the engineering service in a timely manner does not seem like an unreasonable request by the owner and should not be a basis for rejecting the contract. However, the engineer may want to include language which provides protection from service delays caused by conditions outside of the engineer's control.

51. **A** The three principal contractual organization systems used in construction are 1) general contract, 2) construction management, and 3) design

build. In the "general contract" the consulting engineer will be hired by the owner to develop the design and contract documents, as well as to assist in the preparation of the bid documents and provide contract administrative services during the construction phase. The contract documents produced by the engineer will form the basis of the owner's agreement with the contractor. Although the engineer will work closely with the contractor during the construction phase, the engineer will not have a contractual line of privity with the contractor under the "general contract" approach. Design/build project delivery is gaining in popularity and involves a single contract with the owner to provide both design and construction services. In design/build agreements, the engineer would be likely to have a contractual agreement with the construction company.

52. **D** There are five essential elements for making a contract binding.

1. The contract must be established for a legal purpose.
2. The contract must be in proper legal form.
3. There must be consideration (an exchange of something of value, i.e., service in exchange for a fee).
4. The parties must have capacity (the parties must be mentally competent, of legal age, and have the power to enter into the contract).
5. There must be an agreement by both the parties.

Oral contracts may be legally binding in some instances, depending on the circumstances and purpose of the contract. Oral contracts may be difficult to enforce, however, and should not be used for engineering and construction agreements.

53. **C** Past precedent setting legal cases decided at the appellate and supreme court levels will frequently be used in settling legal disputes. The body of law developed from these legal decisions is referred to as "common law."

Statute law is the body of law which is developed by governments creating statutes.

Public law refers to the body of law that governs the actions of governments.

Contract law refers to the body of law that governs contracts between parties.

Tort law is the body of law that governs actions for wrongful acts against another party not stemming from a contract breach.

54. **A** There are typically five distinct service phases that an engineer will provide for an owner on a construction project. These phases are as follows:

- Schematic or preliminary design phase (preparation of initial design alternatives for the owner)

- Design development phase (refinement of the design with more detailed design decisions)

- Contract documents phase (preparation of the detailed construction drawings and specifications as well as preparation of project general conditions)

- Bidding and negotiation phase (aiding the owner in soliciting, receiving and evaluating contractors' bids and proposals)

- Contract administration (services conducted during the construction phase)

Contract administrative services are those services provided by the engineer during the construction phase. In this project phase, the engineer is aiding the owner and contractor as they pursue the completion of their contract obligations. The specific contract administrative services provided by the owner during the construction phase of the project should be defined by the owner/engineer agreement and could entail processing applications for payment and change orders, field observation, project reporting, attending project meetings, clarifying the contract documents, rendering decisions for owner/contractor disputes, and other administrative duties.

55. **A**

56. **C**

57. **D**

58. **A**   The required integral is

$$V = \int_0^{48} p(x)dx$$

There are four intervals within the range (i.e., 1 to 12, 12 to 24, etc.), so

$$n = 4 \quad \text{and} \quad \Delta x = \frac{b-a}{n} = 12$$

The trapezoidal rule yields

$$V = \frac{\Delta x}{2}[f(0) + 2f(1) + 2f(2) + 2f(3) + f(4)]$$
$$= \frac{12}{2}(5 + 2 \times 11 + 2 \times 18 + 2 \times 14 + 10) = 606 \text{ lb}$$

59. **B**   Simpson's rule provides a better estimate:

$$V = \frac{\Delta x}{3}[f(0) + 4f(1) + 2f(2) + 4f(3) + f(4)]$$
$$= \frac{12}{3}(5 + 4 \times 11 + 2 \times 18 + 4 \times 14 + 10) = 604 \text{ lb}$$

60. **D** The derivative of $P(x)$ is $P'(x) = 3x^2 - 8x$. The formula for providing a next estimate, given an initial estimate a_1^1, is

$$a_1^2 = a_1^1 - \frac{P(a_1^1)}{P'(a_1^1)} = -1 - \frac{-1^3 - 4(-1)^2 + 4}{3(-1)^2 - 8(-1)} = -0.909$$

$$a_1^3 = a_1^2 - \frac{P(a_1^2)}{P'(a_1^2)} = -0.909 - \frac{-0.909^3 - 4(-0.909)^2 + 4}{3(-0.909)^2 - 8(-0.909)} = -0.903$$

Note: The exponent on a_i^j represents the iteration number.

Equation Summaries

Appendix A

The following pages are Equation Summary Sheets of the FE/EIT subjects which rely most on equations. They are intended to be used for quick overviews, handy problem solving—and as part of a special strategy for preparing for the exam in its new format.

The selected equations are presented in the same format and nomenclature as is to be found in the NCEES Reference Handbook, a newsprint booklet which is given to all exam applicants. Be prepared! You may find some of the nomenclature to be different from what you are used to using, as the NCEES has apparently obtained some of it from older or obscure texts.

These summaries are useful for study

The current edition of the NCEES Reference Handbook has many extraneous equations which will in all likelihood be of little aid in solving exam problems. As discussed earlier, an excellent exam preparation strategy is to identify the equations you find to be most useful, then highlight them and become acquainted with their position in the Handbook, so that you may quickly access them in the "clean" Handbook during the exam itself. This strategy should maximize your ability to perform as well as possible, given the present exam format.

They can help you anticipate and prepare for NCEES obstacles

A programmable calculator may be pre-programmed to solve many of the problems on the FE exam, e.g., problems involving matrices, permutations, standard deviations, Mohr's circle problems, etc. Or you may choose to buy the HP 48G/GX, which has hundreds of equations and constants preprogrammed. (We offer this superior calculator at a significant discount.) Not all states allow the use of the HP 48G/GX, so be sure you check with your state board.

Calculators are a good way to access equations during the exam

Mathematics
— Selected Equations from the NCEES Reference Handbook —

Straight Line: $y = mx + b$ (slope - intercept form) $m = \dfrac{y_2 - y_1}{x_2 - x_1}$ (slope)

$y - y_1 = m(x - x_1)$ (point - slope form) $m_1 = -\dfrac{1}{m_2}$ (two perpendicular lines)

Quadratic Equation: $ax^2 + bx + c = 0$ $\text{roots} = \dfrac{-b \pm \sqrt{b^2 - 4ac}}{2a}$

Conic Sections:

	General Form	$h = k = 0$
Parabola:	$(y - k)^2 = 2p(x - h)$	$y^2 = 2px$ Focus: $(p/2, 0)$ Directrix: $x = -p/2$
Ellipse:	$\dfrac{(x-h)^2}{a^2} + \dfrac{(y-k)^2}{b^2} = 1$	$\dfrac{x^2}{a^2} + \dfrac{y^2}{b^2} = 1$ Focus: $\left(\sqrt{a^2 - b^2},\ 0\right)$
Hyperbola:	$\dfrac{(x-h)^2}{a^2} - \dfrac{(y-k)^2}{b^2} = 1$	$\dfrac{x^2}{a^2} - \dfrac{y^2}{b^2} = 1$ Focus: $\left(\sqrt{a^2 + b^2},\ 0\right)$
Circle:	$(x - h)^2 + (y - k)^2 = r^2$	$x^2 + y^2 = r^2$

Logarithms:

$\ln x = 2.3026 \log x$ $\qquad \log xy = \log x + \log y \qquad \log x/y = \log x - \log y$

$\log_b b^n = n \qquad\qquad\qquad \log_b b = 1$

$\log x^c = c \log x \qquad\qquad\ \log 1 = 0 \qquad\qquad$ If $b^c = x$, then $\log_b x = c$

Trigonometry:

$\sin\theta = y/r \qquad\qquad \cos\theta = x/r$

$\tan\theta = y/x \qquad\qquad \cot\theta = x/y$

$\csc\theta = r/y \qquad\qquad \sec\theta = r/x$

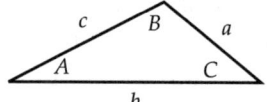

Law of Sines: $\dfrac{a}{\sin A} = \dfrac{b}{\sin B} = \dfrac{c}{\sin C}$

Law of Cosines: $a^2 = b^2 + c^2 - 2bc \cos A$

Identities:

$\tan\theta = \sin\theta/\cos\theta \qquad\qquad\qquad\qquad\quad \sin 2\alpha = 2\sin\alpha\cos\alpha$

$\sin^2\theta + \cos^2\theta = 1 \qquad\qquad\qquad\qquad\quad\ \cos 2\alpha = \cos^2\alpha - \sin^2\alpha$

$\sin(\alpha + \beta) = \sin\alpha\cos\beta + \cos\alpha\sin\beta \qquad\qquad\quad = 2\cos^2\alpha - 1$

$\cos(\alpha + \beta) = \cos\alpha\cos\beta - \sin\alpha\sin\beta \qquad\qquad\quad = 1 - 2\sin^2\alpha$

Appendix A / Equation Summaries—*Mathematics*

Complex Numbers:

$$i = \sqrt{-1} \qquad x + iy = re^{i\theta} \qquad \cos\theta = \frac{e^{i\theta} + e^{-i\theta}}{2}$$

$$r = \sqrt{x^2 + y^2} \qquad e^{i\theta} = \cos\theta + i\sin\theta \qquad \sin\theta = \frac{e^{i\theta} - e^{-i\theta}}{2i}$$

$$(x + iy)^n = r^n(\cos n\theta + i\sin n\theta)$$

Matrices:

Transpose: $\quad \mathbf{B} = \mathbf{A}^T \quad \text{if} \quad b_{ji} = a_{ij}$

Inverse: $\quad \mathbf{A}^{-1} = \dfrac{\text{adj}(\mathbf{A})}{|\mathbf{A}|}$

Adjoint: $\quad \text{adj}(\mathbf{A}) = $ matrix formed by replacing $\mathbf{A}^T$ elements with their cofactors

Cofactor: $\quad \text{cofactor} = \text{minor} \times (-1)^{h+k} \quad$ where $h = $ column, $k = $ row

Minor: $\quad$ minor = determinant that remains after the common row and column are struck out

Vectors:

$$\mathbf{A} \cdot \mathbf{B} = a_x b_x + a_y b_y + a_z b_z \qquad \mathbf{A} \times \mathbf{B} = \begin{vmatrix} \mathbf{i} & \mathbf{j} & \mathbf{k} \\ a_x & a_y & a_z \\ b_x & b_y & b_z \end{vmatrix}$$

$$= |\mathbf{A}||\mathbf{B}|\cos\theta = \mathbf{B} \cdot \mathbf{A}$$

$$= |\mathbf{A}||\mathbf{B}|\mathbf{n}\sin\theta = -\mathbf{B} \times \mathbf{A} \quad \text{where } \mathbf{n} \text{ is } \perp \text{ plane of } \mathbf{A} \text{ and } \mathbf{B}$$

$$\mathbf{i} \cdot \mathbf{i} = \mathbf{j} \cdot \mathbf{j} = \mathbf{k} \cdot \mathbf{k} = 1 \qquad \mathbf{i} \times \mathbf{j} = \mathbf{k}, \quad \mathbf{j} \times \mathbf{k} = \mathbf{i}, \quad \mathbf{k} \times \mathbf{i} = \mathbf{j}$$

Taylor Series:

$$f(x) = f(a) + \frac{f'(a)}{1!}(x - a) + \frac{f''(a)}{2!}(x - a)^2 + \cdots$$

Maclaurin Series: a Taylor series with $a = 0$

Probability and Statistics:

$$P(n, r) = \frac{n!}{(n - r)!} \qquad \text{(permutation of } n \text{ things taken } r \text{ at a time)}$$

$$C(n, r) = \frac{P(n, r)}{r!} = \frac{n!}{r!(n - r)!} \qquad \text{(combination of } n \text{ things taken } r \text{ at a time)}$$

$$\bar{x} = \frac{x_1 + x_2 + \cdots + x_n}{n} \qquad \text{(arithmetic mean)}$$

$$\sigma^2 = \frac{\sum (x_i - \bar{x})^2}{n - 1} \qquad \text{(variance)}$$

$$\sigma = \sqrt{\text{variance}} \qquad \text{(sample standard deviation)}$$

$$\text{median} = \begin{cases} \text{middle value if odd number of items} \\ \tfrac{1}{2}(\text{sum of middle two values}) \text{ if even number of items} \end{cases}$$

mode = value that occurs most often

Calculus: $f'(x) = 0 \begin{cases} \text{maximum} & \text{if } f''(x) < 0 \\ \text{minimum} & \text{if } f''(x) > 0 \end{cases}$

L'Hospital's Rule: $\lim_{x \to a} \dfrac{f(x)}{g(x)} = \lim_{x \to a} \dfrac{f'(x)}{g'(x)}$ if $\dfrac{f(a)}{g(a)} = \dfrac{0}{0}$ or $\dfrac{\infty}{\infty}$

$\dfrac{d}{dx}(uv) = u\dfrac{dv}{dx} + v\dfrac{du}{dx}$ $\dfrac{d}{dx}(\ln u) = \dfrac{1}{u}\dfrac{du}{dx}$ $\dfrac{d}{dx}(\sin u) = \cos u \dfrac{du}{dx}$

$\dfrac{d}{dx}\left(\dfrac{u}{v}\right) = \dfrac{v\,du/dx - u\,dv/dx}{v^2}$ $\dfrac{d}{dx}(e^u) = e^u\dfrac{du}{dx}$ $\dfrac{d}{dx}(\cos u) = -\sin u \dfrac{du}{dx}$

$\dfrac{d}{dx}(u^n) = nu^{n-1}\dfrac{du}{dx}$

$\int x^n dx = \dfrac{x^{n+1}}{n+1} \quad n \neq -1$ $\int \sin x\, dx = -\cos x$ $\int \sin^2 x\, dx = \dfrac{x}{2} - \dfrac{\sin 2x}{4}$

$\int \dfrac{dx}{ax+b} = \dfrac{1}{a}\ln|ax+b|$ $\int \cos x\, dx = \sin x$ $\int \cos^2 x\, dx = \dfrac{x}{2} + \dfrac{\sin 2x}{4}$

$\int e^{ax} dx = \dfrac{1}{a}e^{ax}$

Differential Equations: $y'' + 2ay' + by = f(x)$ (linear, 2nd order, constant coefficient, nonhomogeneous)

Homogeneous solution: $y_h(x) = C_1 e^{r_1 x} + C_2 e^{r_2 x}$ if $r_1 \neq r_2$ where $r^2 + 2ar + b = 0$

$ = (C_1 + C_2 x)e^{r_1 x}$ if $r_1 = r_2$

$ = e^{-ax}(C_1 \cos \beta x + C_2 \sin \beta x)$ if $a^2 < b$. $\beta = \sqrt{b - a^2}$

Particular solution: $y_p = B$ if $f(x) = A$

$ = Be^{\alpha x}$ if $f(x) = Ae^{\alpha x}$

$ = B_1 \sin \omega x + B_2 \cos \omega x$ if $f(x) = A_1 \sin \omega x + A_2 \cos \omega x$

General solution: $y(x) = y_h(x) + y_p(x)$

Mechanics of Materials
—Selected equations from the NCEES Reference Handbook—

Definitions:
$\sigma = \varepsilon E$

$\tau = \gamma G$

$E = 2G(1+v)$

$v = -\dfrac{\varepsilon_{lateral}}{\varepsilon_{longitudinal}}$

E = modulus of elasticity

G = shear modulus

σ and τ = normal and shear stress

ε and γ = normal and shear strain

v = Poisson's ratio

Uniaxial Loading:
$\left.\begin{array}{l}\sigma = \dfrac{P}{A}\\ \varepsilon = \dfrac{\delta}{L}\end{array}\right\}$ $\delta = \dfrac{PL}{AE}$

Thermal Deformation:
$\delta_t = \alpha L(T - T_o)$

α = coefficient of thermal expansion

Thin-walled Pressure Vessel:
$\sigma_t = \dfrac{pD}{2t}$ hoop (circumferential) stress

$\sigma_a = \dfrac{pD}{4t}$ axial (longitudinal) stress

t = cylinder thickness
D = cylinder diameter
p = pressure

Stress and Strain:

Stress Condition

Mohr's Circle

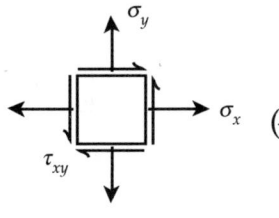

Maximum and Minimum Stresses

$\sigma_1 = \sigma_{max} = \dfrac{\sigma_x + \sigma_y}{2} + \left[(\sigma_x - \sigma_y)^2/4 + \tau_{xy}^2\right]^{1/2}$

$\sigma_2 = \sigma_{min} = \dfrac{\sigma_x + \sigma_y}{2} - \left[(\sigma_x - \sigma_y)^2/4 + \tau_{xy}^2\right]^{1/2}$

$\tau_{max} = \dfrac{\sigma_1 - \sigma_2}{2}$ = radius of Mohr's circle

3-D Strain: $\varepsilon_x = \dfrac{1}{E}\left[\sigma_x - v(\sigma_y + \sigma_z)\right]$

$\gamma_{xy} = \dfrac{\tau_{xy}}{G}$

Torsion: $\tau = \dfrac{Tr}{J}$ (shear stress)

$\phi = \dfrac{TL}{JG}$ (angle of twist)

J = polar moment of inertia
$= \pi r^4/2$ for a circle

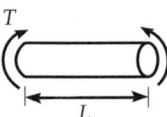

Beams: $V = \dfrac{dM}{dx}$

$\sigma = -\dfrac{My}{I}$

$\tau = \dfrac{VQ}{Ib}$

$EIy'' = M$

V = vertical shear force

I = centroidal moment of inertia
$= bh^3/12$ for a rectangle
$= \pi r^4/4$ for a circle

Q = moment of area between
y - position and top or bottom

differential equation of deflection curve

M = bending moment

y = distance from neutral axis

Columns: $P_{cr} = \dfrac{\pi^2 EI}{k^2 L^2}$ $k = \begin{cases} 1 & \text{ends pinned} \\ 0.5 & \text{ends fixed} \\ 0.7 & \text{one pinned, one fixed} \\ 2 & \text{one fixed, one free} \end{cases}$

Dynamics
—Selected equations from the NCEES Reference Handbook—

Kinematics (motion only)

Tangential and Normal Components: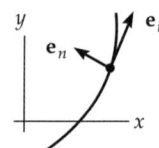

$$\mathbf{a} = \frac{dv_t}{dt}\mathbf{e_t} + \frac{v_t^2}{\rho}\mathbf{e_n}$$

$$\mathbf{v} = v_t\mathbf{e_t}$$

ρ = radius of curvature

Plane Circular Motion: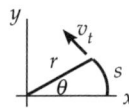

$\mathbf{e_r} = -\mathbf{e_n}$

$\mathbf{e_\theta} = \mathbf{e_t}$

$\omega = \dot{\theta} = \dfrac{v_t}{r}$

$\alpha = \dot{\omega} = \ddot{\theta} = \dfrac{a_t}{r}$

$v_t = r\omega$

$a_t = r\alpha$

$a_n = \dfrac{v_t^2}{r} = r\omega^2$

$s = r\theta$

Straight Line Motion:

$s = s_o + v_o t + a_o t^2/2$

$v = v_o + a_o t$

$v^2 = v_o^2 + 2a_o(s - s_o)$

Projectile Motion: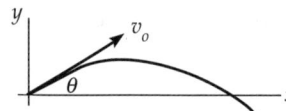

$a_x = 0, \quad a_y = -g$

$v_x = v_o \cos\theta$

$v_y = v_o \sin\theta - gt$

$x = v_o t \cos\theta$

$y = v_o t \sin\theta - \tfrac{1}{2}gt^2$

Kinematics (forces and motion)

$$\sum \mathbf{F} = \frac{d}{dt}(m\mathbf{v}), \quad \sum F_t = ma_t = m\frac{dv_t}{dt}, \quad \sum F_n = ma_n = m\frac{v_t^2}{\rho}$$

Impulse and Momentum:

$$m[v_x(t) - v_x(0)] = \int_0^t F_x(t)dt \quad \text{or} \quad \text{change in momentum} = \text{impulse}$$

Work and Energy:

$PE_1 + KE_1 + W_{1\rightarrow 2} = PE_2 + KE_2$

where

$KE = \tfrac{1}{2}mv^2$

$PE = mgh$ (gravity)

$= \tfrac{1}{2}kx^2$ (spring)

$W_{1\rightarrow 2}$ = friction force work

Impact:

$m_1 v_1 + m_2 v_2 = m_1 v_1' + m_2 v_2'$

$e = -\dfrac{v_{1n}' - v_{2n}'}{v_{1n} - v_{2n}} = \begin{cases} 1 & \text{elastic} \\ 0 & \text{plastic} \end{cases}$

v_1, v_2 = velocities before impact

v_1', v_2' = velocities after impact

Rotation:

$I_o \alpha = \sum M_o$ where $I_o = \int (x^2 + y^2)dm$, rotation about O.

constant M: $\alpha = \dfrac{M}{I}$

$\omega = \omega_o + \dfrac{M}{I}t$

$\theta = \theta_o + \omega_o t + \dfrac{M}{2I}t^2$

work and energy: $I_o \dfrac{\omega^2}{2} - I_o \dfrac{\omega_o^2}{2} = \int_{\theta_o}^{\theta} M d\theta$

Banking of Curves: $\tan\theta = \dfrac{v^2}{rg}$ where r = radius of curvature

θ = angle between surface and horizontal

Electric Circuits
—Selected Equations from the NCEES Reference Handbook—

Electrostatics:

$F_2 = \dfrac{Q_1 Q_2}{4\pi\varepsilon r^2}$ (force on charge 2 due to charge 1) $\quad \varepsilon = $ permittivity — $C^2/N \cdot m^2 = F/m$
$\qquad\qquad\qquad\qquad\qquad\qquad\qquad\qquad\qquad\qquad\qquad = 8.85 \times 10^{-12}$ for air or free space

$E = \dfrac{Q}{4\pi\varepsilon r^2}$ (electric field intensity due to point charge Q — C)

$E_L = \dfrac{\rho_L}{2\pi\varepsilon r}$ (radial field due to line charge ρ_L — C/m)

$E_s = \dfrac{\rho_s}{2\varepsilon}$ (plane field due to sheet charge ρ_s — C/m^2)

$Q = \oint \varepsilon \mathbf{E} \cdot d\mathbf{A}$ (enclosed charge — C)

$E = \dfrac{V}{d}$ (electric field between plates with potential difference V separated by the distance d)

$H = \dfrac{I}{2\pi r}$ (magnetic field strength due to current in long wire)

$B = \mu H$ (magnetic flux density)

$\mathbf{F} = I\mathbf{L} \times \mathbf{B}$ (force on conductor) $\qquad \mathbf{L}$ = length vector of conductor

DC Circuits:

Resistors: $\quad V = IR \quad$ (Ohm's law) $\qquad R_T = R_1 + R_2 + \cdots \quad$ (series)

$P = VI = \dfrac{V^2}{R} = I^2 R \quad$ (power) $\qquad R_T = \left[\dfrac{1}{R_1} + \dfrac{1}{R_2} + \cdots\right]^{-1} \quad$ (parallel)

Capacitors: $\quad i = C\dfrac{dv}{dt} \qquad$ energy stored $= \tfrac{1}{2}Cv^2 \qquad C_{eq} = C_1 + C_2 + \cdots \quad$ (parallel)

$v = \dfrac{1}{C}\int i\, dt \qquad\qquad\qquad\qquad\quad C_{eq} = \left[\dfrac{1}{C_1} + \dfrac{1}{C_2} + \cdots\right]^{-1} \quad$ (series)

Inductors: $\quad i = \dfrac{1}{L}\int v\, dt \qquad$ energy stored $= \tfrac{1}{2}Li^2 \qquad L_{eq} = L_1 + L_2 + \cdots \quad$ (series)

$v = L\dfrac{di}{dt} \qquad\qquad\qquad\qquad\qquad L_{eq} = \left[\dfrac{1}{L_1} + \dfrac{1}{L_2} + \cdots\right]^{-1} \quad$ (parallel)

Kirchhoff Voltage Law (KVL): $\quad \sum V_{rises} = \sum V_{drops} = 0$

Kirchhoff Current Law (KCL): $\quad \sum I_{in} = \sum I_{out}$

Thevenin equivalent circuit:

$R_{eq} = \dfrac{V_{eq}}{I_{sc}}$

I_{sc} = short circuit current
V_{eq} = open circuit voltage

RC Transients:

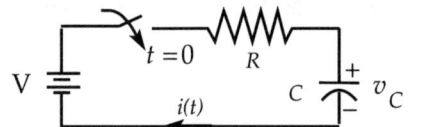

$$v_C(t) = v_C(0)e^{-t/RC} + V\left(1 - e^{-t/RC}\right)$$
$$i(t) = \{[V - v_C(0)]/R\}e^{-t/RC}$$

RL Transients:

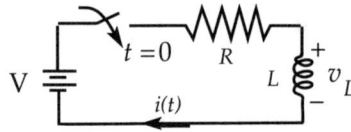

$$v_L(t) = -i(0)Re^{-Rt/L} + Ve^{-Rt/L}$$
$$i(t) = i(0)e^{-Rt/L} + \frac{V}{R}\left(1 - e^{-Rt/L}\right)$$

Operational Amplifiers:

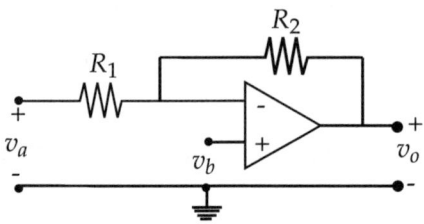

$$v_o = -\frac{R_2}{R_1}v_a + \left(1 + \frac{R_2}{R_1}\right)v_b$$

inverting if $v_b = 0$
non-inverting if $v_a = 0$

AC Circuits: $\quad f = \dfrac{1}{T} = \dfrac{\omega}{2\pi}$
(single phase)

f = frequency (Hz)
T = period (sec)
ω = angular frequency (rad/s)

$V_{avg} = \dfrac{2}{\pi}V_{max}$ \quad (full-wave rectified sine wave)

$V_{avg} = \dfrac{1}{\pi}V_{max}$ \quad (half-wave rectified sine wave)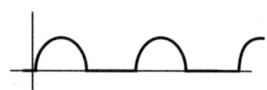

$V_{rms} = \dfrac{1}{\sqrt{2}}V_{max}$ \quad (full-wave rectified sine wave)

$V_{rms} = \dfrac{1}{2}V_{max}$ \quad (half-wave rectified sine wave)

Resistor: $\quad Z = R \quad\quad\quad\quad Z$ = Impedance

Capacitor: $\quad Z = -\dfrac{j}{\omega C} = -jX \quad\quad X$ = Reactance

Inductor: $\quad Z = j\omega L = jX$

$V = IZ$

$P = \tfrac{1}{2}V_{max}I_{max}\cos\theta = V_{rms}I_{rms}\cos\theta$ \quad (real power) \quad ($\theta = 0$ for resistors)

$Q = \tfrac{1}{2}V_{max}I_{max}\sin\theta = V_{rms}I_{rms}\sin\theta$ \quad (reactive power)

p.f. = $\cos\theta$ \quad (power factor)

Resonance: $\quad f = \dfrac{1}{2\pi\sqrt{LC}}$ \quad (resonant frequency for series and parallel circuits)

Fluid Mechanics
—Selected Equations from the NCEES Reference Handbook—

Properties:

$\rho = \dfrac{m}{V}$ (density) $\qquad$ $\tau_n = -p$ (normal stress)

$\gamma = \rho g$ (specific weight) $\qquad$ $\tau_t = \mu \dfrac{dv}{dy}$ (tangential stress)

$\nu = \dfrac{\mu}{\rho}$ (kinematic viscosity) $\qquad$ μ = dynamic viscosity

Statics:

$p_2 - p_1 = -\gamma h$ $\qquad$ (h is vertical upward) $\qquad$ $F = \gamma h_C A$

$F_{\text{buoyant}} = \gamma V_{\text{displaced}}$ (Archimedes' principle) $\qquad$ $z^* = \dfrac{I_C}{A Z_C}$

One-Dimensional Flows:

$A_1 V_1 = A_2 V_2$ $\qquad$ (continuity equation)

$Q = AV$ $\qquad$ (flow rate)

$\dot{m} = \rho A V$ $\qquad$ (mass flow rate)

$-\dfrac{\dot{W_s}}{\gamma Q} + \dfrac{p_1}{\gamma} + \dfrac{V_1^2}{2g} + z_1 = \dfrac{p_2}{\gamma} + \dfrac{V_2^2}{2g} + z_2 + h_f$ $\qquad$ (Energy Equation — if $h_f = \dot{W}_s = 0$, then Bernoulli Eq.)

$h_f = f \dfrac{L}{D} \dfrac{V^2}{2g}$ $\qquad$ (Darcy's Equation — find f on Moody Diagram)

$\text{Re} = \dfrac{VD\rho}{\mu}$ $\qquad$ (Reynolds Number)

$h_{f,\text{ fitting}} = C \dfrac{V^2}{2g}$ $\qquad$ (minor losses — C is loss coefficient)

$\sum \mathbf{F} = \rho Q (\mathbf{V_2} - \mathbf{V_1})$ $\qquad$ (Momentum equation)

Perfect Gas:

$p = \rho R T$ $\qquad$ (perfect gas law)

$c = \sqrt{kRT}$ $\qquad$ (speed of sound)

$\text{M} = \dfrac{V}{c}$ $\qquad$ (Mach number)

Similitude: If viscous effects dominate (internal flows) then Reynolds numbers on prototype and model are equated:

$$(\text{Re})_p = (\text{Re})_m \quad \text{or} \quad \left(\dfrac{V \ell \rho}{\mu}\right)_p = \left(\dfrac{V \ell \rho}{\mu}\right)_m$$

If gravity dominates (dams, weirs, ships) then Froude numbers are equated:

$$(\text{Fr})_p = (\text{Fr})_m \quad \text{or} \quad \left(\dfrac{V^2}{\ell g}\right)_p = \left(\dfrac{V^2}{\ell g}\right)_m$$

Open Channel:

$Q = \dfrac{C}{n} A R^{2/3} S^{1/2}$ $\qquad$ where $\qquad$ $R = \dfrac{A}{P_{\text{wetted}}}$

$C = \begin{cases} 1.0 & \text{metric} \\ 1.49 & \text{english} \end{cases}$

Thermodynamics
—Selected Equations from the NCEES Reference Handbook—

Properties:
- P (absolute pressure, kPa or lbf/in^2)
- $v = \dfrac{V}{m}$ (specific volume, m^3/kg or ft^3/lbm)
- u (internal energy, kJ/kg or Btu/lbm)
- $h = u + Pv$ (enthalpy, kJ/kg or Btu/lbm)
- s (entropy, kJ/kg·K or Btu/lbm-°R)
- c_p (constant pressure specific heat, kJ/kg·K or Btu/lbm-°R)
- c_v (constant volume specific heat, kJ/kg·K or Btu/lbm-°R)
- $x = \dfrac{m_v}{m_{total}}$ (quality)

Two phase system:
$$v = v_f + x v_{fg} \quad \text{where} \quad v_{fg} = v_g - v_f$$
$$h = h_f + x h_{fg}$$
v_f = saturated liquid value
v_g = saturated vapor value

Ideal gas:
$$Pv = RT, \quad PV = mRT \quad \text{where} \quad R = \frac{\overline{R}}{M}, \quad \overline{R} = 8.314 \frac{\text{kJ}}{\text{kmol} \cdot \text{K}} \quad \text{or} \quad 1545 \frac{\text{ft-lbf}}{\text{lbmol-°R}}$$
$$\Delta u = c_v \Delta T, \quad \Delta h = c_p \Delta T$$
$$\Delta s = c_p \ln \frac{T_2}{T_1} - R \ln \frac{P_2}{P_1} = c_v \ln \frac{T_2}{T_1} + R \ln \frac{v_2}{v_1}$$
$$\left. \frac{T_2}{T_1} = \left(\frac{P_2}{P_1} \right)^{\frac{k-1}{k}} = \left(\frac{v_1}{v_2} \right)^{k-1}, \quad \begin{array}{c} P_2 v_2^k = P_1 v_1^k \\ k = c_p/c_v \end{array} \right\} \text{(constant entropy process)}$$

First law (system):
$$q - w = \Delta u \quad \text{where} \quad w = \int P\, dv$$
$$= RT \ln \frac{v_2}{v_1} = RT \ln \frac{P_1}{P_2} \quad \text{(isothermal process with ideal gas)}$$

First law (control volume):
- $h_i + V_i^2/2 = h_e + V_e^2/2$ (nozzles, diffusers)
- $h_i = h_e + w$ (turbine, compressor)
- $h_i = h_e$ (throttling device, valve)
- $h_i + q = h_e$ (boilers, condensers, evaporators)

i = inlet
e = exit

Cycles:
$$\eta = \frac{W}{Q_H} = \frac{Q_H - Q_L}{Q_H} \quad \text{(efficiency)} \qquad \text{COP} = \frac{Q_H}{W} \quad \text{(heat pump)}$$
$$= 1 - \frac{T_L}{T_H} \quad \text{(Carnot cycle)} \qquad\qquad = \frac{Q_L}{W} \quad \text{(refrigerator)}$$

Second Law: No engine can produce work while transferring heat with a single reservoir. (Kelvin-Planck)

No refrigerator can operate without a work input. (Clausius)

$$\Delta S \geq \int \frac{\delta Q}{T} \qquad \Delta S = \frac{Q}{T} \quad \text{(reservoir or } T = \text{const)}$$
$$\Delta S_{total} = \Delta S_{surr} + \Delta S_{system} \geq 0 \qquad \Delta S = C_p \ln \frac{T_2}{T_1} \quad \text{(solid or liquid)}$$

Heat Transfer:
$$q = -kA \frac{dT}{dx} \quad \text{(conduction)} \qquad k = \text{conductivity}$$
$$= -kA \frac{T_2 - T_1}{L} \quad \text{(through a wall)} \qquad R = \frac{L}{kA} \quad \text{(resistance factor)}$$
$$q = hA(T_1 - T_2) \quad \text{(convection)} \qquad R = \frac{1}{hA} \quad \text{(resistance factor)}$$
$$= \varepsilon \sigma A \left(T_1^4 - T_2^4\right) F_{12} \quad \text{(radiation)} \qquad h = \text{convection coefficient}$$

$\varepsilon = 1$ for black body (emissivity)

$\sigma = 5.67 \times 10^{-8} \dfrac{W}{m^2 \cdot K^4}$ (Stefan - Boltzmann constant)

$F_{12} = 1$ if one body encloses the other (shape factor)

English and SI Units

Appendix B

The following tables present the SI (Systems International) units and the conversion of English units to SI units, along with some of the more common conversion factors.

SI Prefixes

Multiplication Factor	Prefix	Symbol
10^{15}	peta	P
10^{12}	tera	T
10^{9}	giga	G
10^{6}	mega	M
10^{3}	kilo	k
10^{-1}	deci	d
10^{-2}	centi	c
10^{-3}	mili	m
10^{-6}	micro	μ
10^{-9}	nano	n
10^{-12}	pico	p
10^{-15}	femto	f

SI Base Units

Quantity	Name	Symbol
length	meter	m
mass	kilogram	kg
time	second	s
electric current	ampere	A
temperature	kelvin	K
amount of substance	mole	mol
luminous intensity	candela	cd

SI Derived Units

Quantity	Name	Symbol	In Terms of Other Units
area	square meter		m^2
volume	cubic meter		m^3
velocity	meter per second		m/s
acceleration	meter per second squared		m/s^2
density	kilogram per cubic meter		kg/m^3
specific volume	cubic meter per kilogram		m^3/kg
frequency	hertz	Hz	s^{-1}
force	newton	N	$m \cdot kg/s^2$
pressure, stress	pascal	Pa	$kg/(m \cdot s^2)$
energy, work, heat	joule	J	$N \cdot m$
power	watt	W	J/s
electric charge	coulomb	C	$A \cdot s$
electric potential	volt	V	W/A
capacitance	farad	F	C/V
electric resistance	ohm	Ω	V/A
conductance	siemens	S	A/V
magnetic flux	weber	Wb	$V \cdot s$
inductance	henry	H	Wb/A
viscosity	pascal second		$Pa \cdot s$
moment (torque)	meter newton		$N \cdot m$
heat flux	watt per square meter		W/m^2
entropy	joule per kelvin		J/K
specific heat	joule per kilogram-kelvin		$J/(kg \cdot K)$
conductivity	watt per meter-kelvin		$W/(m \cdot K)$

Conversion Factors to SI Units

English	SI	SI Symbol	To Convert from English to SI Multiply by
Area			
square inch	square centimeter	cm2	6.452
square foot	square meter	m2	0.09290
acre	hectare	ha	0.4047
Length			
inch	centimeter	cm	2.54
foot	meter	m	0.3048
mile	kilometer	km	1.6093
Volume			
cubic inch	cubic centimeter	cm^3	16.387
cubic foot	cubic meter	m^3	0.02832
gallon	cubic meter	m^3	0.003785
gallon	liter	L	3.785
Mass			
pound mass	kilogram	kg	0.4536
slug	kilogram	kg	14.59
Force			
pound	newton	N	4.448
kip(1000 lb)	newton	N	4448
Density			
pound/cubic foot	kilogram/cubic meter	kg/m^3	16.02
pound/cubic foot	grams/liter	g/L	16.02
Work, Energy, Heat			
foot-pound	joule	J	1.356
Btu	joule	J	1055
Btu	kilowatt-hour	kWh	0.000293
therm	kilowatt-hour	kWh	29.3

Conversion Factors to SI Units (continued)

English	SI	SI Symbol	To Convert from English to SI Multiply by
Power, Heat, Rate			
horsepower	watt	W	745.7
foot pound/sec	watt	W	1.356
Btu/hour	watt	W	0.2931
Btu/hour-ft^2-°F	watt/meter squared-°C	W/m$^2 \cdot$ °C	5.678
tons of refrig.	kilowatts	kW	3.517
Pressure			
pound/square inch	kilopascal	kPa	6.895
pound/square foot	kilopascal	kPa	0.04788
inches of H$_2$O	kilopascal	kPa	0.2486
inches of Hg	kilopascal	kPa	3.374
one atmosphere	kilopascal	kPa	101.3
Temperature			
Fahrenheit	Celsius	°C	5 (°F − 32)/9
Fahrenheit	kelvin	K	5 (°F + 460)/9
Velocity			
foot/second	meter/second	m/s	0.3048
mile/hour	meter/second	m/s	0.4470
mile/hour	kilometer/hour	km/h	1.609
Acceleration			
foot/second squared	meter/second squared	m/s^2	0.3048
Torque			
pound-foot	newton-meter	N $\cdot$ m	1.356
pound-inch	newton-meter	N $\cdot$ m	0.1130
Viscosity, Kinematic Viscosity			
pound-sec/square foot	newton-sec/square meter	N $\cdot$ s/m^2	47.88
square foot/second	square meter/second	m^2/s	0.09290
Flow Rate			
cubic foot/minute	cubic meter/second	m^3/s	0.0004719
cubic foot/minute	liter/second	L/s	0.4719
Frequency			
cycles/second	hertz	Hz	1.00

Conversion Factors

Length
1 cm = 0.3937 in
1 m = 3.281 ft
1 yd = 3 ft
1 mi = 5280 ft
1 mi = 1760 yd
1 km = 3281 ft

Area
1 cm^2 = 0.155 in^2
1 m^2 = 10.76 ft^2
1 ha = 10^4 m^2
1 acre = 100 m^2
1 acre = 4047 m^2
1 acre = 43,560 ft^2
1 acre-ft = 43,560 ft^3
1 m^3 = 1000 L

Volume
1 ft^3 = 28.32 L
1 L = 0.03531 ft^3
1 L = 0.2642 gal
1 m^3 = 264.2 gal
1 ft^3 = 7.481 gal
1 m^3 = 35.31 ft^3

Velocity
1 m/s = 3.281 ft/s
1 mph = 1.467 ft/s
1 mph = 0.8684 knot
1 knot = 1.688 ft/s
1 km/h = 0.2778 m/s
1 km/h = 0.6214 mph

Force
1 lb = 4.448 x 10^5 dyne
1 lb = 32.17 pdl
1 lb = 0.4536 kg
1 N = 10^5 dyne
1 N = 0.2248 lb
1 kip = 1000 lb

Mass
1 oz = 28.35 g
1 lb = 0.4536 kg
1 kg = 2.205 lb
1 slug = 14.59 kg
1 slug = 32.17 lb

Work and Heat
1 BTU = 778.2 ft-lb
1 BTU = 1055 J
1 Cal = 3.088 ft-lb
1 J = 10^7 ergs
1 kJ = 0.9478 ft-lb
1 BTU = 0.2929 W·hr
1 ton = 12,000 BTU/hr
1 kWh = 3414 BTU
1 quad = 10^{15} BTU
1 therm = 10^5 BTU

Power
1 Hp = 550 ft-lb/s
1 HP = 33,000 ft-lb/min
1 Hp = 0.7067 BTU/s
1 Hp = 2545 BTU/hr
1 Hp = 745.7 W
1 W = 3.414 BTU/hr
1 kW = 1.341 Hp

Volume Flow Rate
1 cfm = 7.481 gal/min
1 cfm = 0.4719 L/s
1 m^3/s = 35.31 ft^3/s
1 m^3/s = 2119 cfm
1 gal/min = 0.1337 cfm

Torque
1 N·m = 10^7 dyne·cm
1 N·m = 0.7376 lb-ft
1 N·m = 10 197 g·cm
1 lb-ft = 1.356 N·m

Viscosity
1 lb-s/ft^2 = 478 poise
1 poise = 1 g/cm·s
1 N·s/m^2 = 0.02089 lb-s/ft^2

Pressure
1 atm = 14.7 psi
1 atm = 29.92 in Hg
1 atm = 33.93 ft H_2O
1 atm = 1.013 bar
1 atm = 1.033 kg/cm^2
1 atm = 101.3 kPa
1 psi = 2.036 in Hg
1 psi = 6.895 kPa
1 psi = 68 950 dyne/cm^2
1 ft H_2O = 0.4331 psi
1 kPa = 0.145 psi

PASS YOUR EXAM WITH THE
ULTRA PE PREP SYSTEM

$25 OFF! YOUR PERSONAL SYSTEM!

Great Lakes Press 5-Step System is the best way to ensure exam readiness!

To create your own "Ultra PE Prep System," simply order a concise review, a past sample exam, a discounted handbook, and be sure you own and know how to use a powerful calculator—and you have the resources you need to pass the PE exam!

Step #1: Get the Best PE Review!
- *Principles & Practice of Civil / Mechanical / Electrical Engineering* edited by Merle C. Potter, PhD, PE, and written by teams of 9–10 veteran professors. These highly effective PE reviews are from the publisher of the best–selling FE/EIT review. Start with a great review…these are the only titles that concisely cover all critical aspects of the PE exam. Hundreds of excellent exam-simulating practice problems. 600-700 pp. (Solutions manuals to problems available separately.)

Step #2: Select a Past Sample Exam!
- *Official NCEES Sample Problems & Solutions*. These include actual past PE exam problems, all fully solved by official scorers. All PE exams are available in the latest printing. After your initial prep, take your sample exam (open book) to verify readiness or identify weaknesses. Use your review, handbook, and calculator to assist you—be sure you have easy access to all necessary information.

Step #3: Order a Handbook!
The PE exam is open–book, so be sure you have a current, all–inclusive handbook for easy, one–stop access to all necessary information.

- *Standard Handbook for Civil Engineers* by F. Merritt (4th ed., McGraw–Hill). New edition of the most thorough compilation of facts and figures. Best guide, best buy, and best seller since 1968!! 1,456 pp. List price: $150, GLP price: $119!
- *Mark's Standard Handbook for Mechanical Engineers* by E. Avalline (10th ed., McGraw–Hill). Practical advice and quick answers on all ME standards and practices. 1,792 pp. List price: $150, GLP price: $119!
- *Standard Handbook for Electrical Engineers* by Fink & Beaty (13th ed., McGraw–Hill). The premier EE reference. Up–to–date info in all areas of EE. 2,000 pp. List price: $125, GLP price: only $99!
- *Perry's Chemical Engineer's Handbook* by R. Perry (7th ed., McGraw–Hill). The best info on all aspects of ChemE…all in one place! The standard for over 40 years. 2,640 pp. List price: $150, GLP price: $119!
- *Structural Engineering Handbook* by Gaylord (4th ed., McGraw–Hill). Up–to–date information on all areas of Structural Engineering. Approx. 2,000 pp. List price: $125, GLP price: only $99!

Step #4: Don't Forget a Test–Beating Calculator!
- *HP 48GX Programmable Calculator*. Most states allow sophisticated calculators to be used when taking PE & FE exams. The HP 48 has hundreds of built–in equations you will need. GX plug–in cards offers exact FE/PE equations! Also, be sure you own *JumpStart the HP 48* (3rd ed.), a handy user's guide written to help engineers get the most out of these Hewlett Packard calculators.

Optional: ASCE Video Review Package (for PE Civil only)
- Newly updated with 23 hours of review from actual ASCE PE course. Includes comprehensive workbook. GLP throws in a free copy of our PE Civil Review and Solutions Manual—$93 value!

Step #5: Fill Out the Quick'n'Easy Order Form Below

(please print)

Name _____

Address _____

City / State / ZIP _____

Phone with Area Code _____

Credit Card: ☐ ☐   Exp. Date _____

Credit Card # _____

30–day Money Back Guarantee on GLP titles only.

Call **800-837-0201** or fax completed form to **636-273-6086**.

Or mail this form with credit card info or check or money order to:

Great Lakes Press
PO Box 550
Wildwood, MO 63040-0550

www.glpbooks.com   custserv@glpbooks.com   *Prices are subject to change.*

Description	Price	Ship	Quan.	Discount	Subtotal
PE CE / ME Review	$79.95 CE / $74.95 ME	$5		Package Savings on the Ultra PE Prep System! Take $25 off if ordering from Steps 1, 2, & 3!	
PE EE Review	$69.95	$5			
PE Solutions CE / ME / EE	$19.95	$3			
NCEES Sample PE Exam CE / ME / EE / ChemE /Enviro	$30	$5			
NCEES PE Exam, Structural	$45	$5			
CE / ME / EE / ChemE / Struct Handbooks	$99/$119	$7			
Calculator HP 48GX	$139.95	$7		N/A	
GX Card for FE / PE	$124.95	$7		N/A	
JumpStart the HP 48G/GX	$21.95	$3		N/A	
PE Civil Video Package	$595	$7		N/A	
Tax (MI & MO residents add 6.0% sales tax)					
TOTAL					

5/2001

CATALOG OF TITLES BY GREAT LAKES PRESS

FREE! Interactive CD With FE Titles!

Editor and Principal Author: Merle C. Potter, Ph.D., P.E.

Fundamentals of Engineering Review/Gen'l
10th edition, $54.95
The most effective review for the FE/EIT exam. Written by nine PhD professors for the A.M. and General P.M. sessions. This is the only review available that contains only what you need to pass—no more, no less! Includes full reviews of all 12 major general subjects; over 1,000 problems and full solutions, two fully–solved practice exams, and NCEES equation summaries. 669 pages. Includes coupon for FREE full-feature CD exams and StudyDirector™!

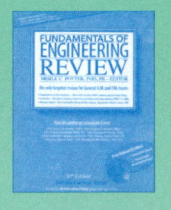

FE/EIT Discipline Reviews, 4 vols., 5th ed, $29.95
Covering all disciplines: CE, ME, EE, ChemE/IE, in four volumes. Each review teaches all essential equations in the *NCEES Reference Handbook*. Each vol. includes review, practice problems, solutions, & practice exam. Written by PhD teaching professors. 140-160 pp. each. Includes coupon for FREE exams CD with StudyDirector™!

NCEES FE Sample Exam & FE Supplied Reference Handbook, Exam $21.95, Handbook $10
Official 141-page sample exam from 2000. The *Reference Handbook* is a must for preparation—it is your only exam site resource and familiarity with it is essential. GLP reviews are keyed to the *Handbook*!

Jump Start the HP 48G/GX, 3rd edition, $21.95
Unique user manual written expressly for engineers. Superior to HP manual. Get the most out of your calculator! Makes passing tests much easier! Covers: Using Equations, Solved Problems, Special Applications, Writing Programs, Basic Statistics, File Transfer, Matrices, and more. 240 pages.

GRE, GMAT & GRE/Engineering Time•Savers™, $19.95 each
The most authoritative guides to these tests—authors are PhD teaching professors from major universities. Study same subjects as in other reviews…in much less time! 3 practice tests with solutions.

FULL REVIEWS FOR THE PRINCIPLES & PRACTICE OF ENGINEERING (PE) EXAMS:

Great Lakes Press' PE Exam reviews are written by 9-10 professors, experts in their fields. Reviews all major areas tested. Hundreds of excellent exam-simulating practice problems.

Principles & Practice of Civil Engineering Review, 4th edition, $79.95
The most concise coverage of all major subject areas tested. 650 pages. (Practice problems solutions manual, $19.95, 160 pp.)

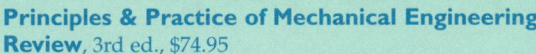

Principles & Practice of Mechanical Engineering Review, 3rd ed., $74.95
The most concise coverage of all major subject areas tested. 586 pages. (Practice problems solutions manual, $19.95, 130 pp.)

Principles & Practice of Electrical Engineering Review, $69.95
The most concise coverage of all major subject areas tested. 397 pages. (Practice problems solutions manual, $19.95, 100 pp.)

Quick'n'Easy Order Form *(please print)*

Name_____

Address_____

City / State / ZIP _____

Phone with Area Code _____

Credit Card: ❑ ❑   Exp. Date _____

Credit Card # _____

30–day Money Back Guarantee on GLP titles only.
Call **800-837-0201** or fax completed form to **636-273-6086**.

Or mail this form with credit card info or check or money order to:
Great Lakes Press
PO Box 550
Wildwood, MO 63040-0550

www.glpbooks.com   custserv@glpbooks.com   *Prices are subject to change.*

Description		Price	Quantity	Subtotal
Fundamentals of Engineering/General		$54.95		
FE/EIT Discipline: CE / ME / EE / ChemE/IE		$29.95		
NCEES: FE Exam / FE Handbook		$21.95/$10		
PE Review: CE / ME	$79.95 CE / $74.95 ME			
PE Review: EE		$69.95		
Solutions: CE / ME / EE		$19.95		
JumpStart HP 48 G/GX		$21.95		
GRE / GMAT Time•Saver		$19.95		
GRE / Engineering Review		$19.95		
Tax (MI & MO residents add 6% sales tax)				
Shipping ($7 first book + $1 ea. add'l)				
❑ Send a Catalog to a Friend or Colleague		**TOTAL**		

Name/Company _____

Address _____

5/2001

GREAT LAKES PRESS

Reader Remarks & Rewards Survey

Your suggestions help us to improve this review continuously. As a way of saying thanks, we'll send you a FREE FE/EIT exams CD ($29.95 value) when you fill out and return this card. (Please include $5 shipping/handling.)

ABOUT YOU

Name _____

Address _____

City / State / ZIP _____

Phone / E-mail _____

Field / Position _____

❏ "You can tell them I said so!"   ❏ "Hey, send me that free CD!"
($5 ship/hndl, incl. check or CC#)

COMMENTS

_____

_____

_____

_____

Any content we missed? _____

_____

QUICK SURVEY

1) Your review?   F/G F/CE F/ME F/EE F/IE F/ChE P/CE P/ME P/EE

2) Overall, this book is:
    A Too sketchy
    B Just about right
    C Too much material

3) The problems in this book are:
    A Too easy
    B Just about right
    C Too difficult

4) The solutions are:
    A Too sketchy
    B Just about right
    C Too much explanation

5) Did you participate in a review course?
    A YES, Course location/name _____
    B NO

6) Did you use other material in your preparation?
Please list _____

7) How long since your undergraduate college graduation:
_____ years   _____ haven't yet

8) Rank factors in order of influence on your initial appraisal of this book (1 being most important):
_____ Price   _____ Reputation
_____ Written by professors   _____ Presentation of material
_____ Depth / Amount of material
_____ Other _____

REWARDS FOR ERRATA!   *(Attach separate sheet if desired.)*

Think you found a mistake? We happily offer up to $3 for each error that has not already been discovered, depending on relevance

Error _____   Proposed Correction _____

Page Number _____   Problem or Example Number _____

SEND INFO TO A FRIEND, COLLEAGUE, OR COMPANY MANAGER

The following people would appreciate a *one-time* mailing of a catalog of your FE & PE resources.

Name _____   Name _____   Name _____

Address _____   Address _____   Address _____

_____   _____   _____

City _____   City _____   City _____

State / ZIP _____   State / ZIP _____   State / ZIP _____

❏ Civil   ❏ Mechanical   ❏ Civil   ❏ Mechanical   ❏ Civil   ❏ Mechanical
❏ Electrical   ❏ Other   ❏ Electrical   ❏ Other   ❏ Electrical   ❏ Other

TO RECEIVE *FREE* CD ...a $29.95 value!...fill out post-paid form, return with credit card info or check payable to "Great Lakes Press," for $5 shipping/handling.

Credit card # _____   Exp. Date: _____

CC Billing Address: _____

Hey! Send me my FREE FE Exams CD!

☐ ✓ Check this box, fill out entire form on reverse side...
...then tear, fold and send in this whole postpaid card!
(please allow 2 weeks for delivery)

• CD includes 6 solved exams, Study-Director™
...and much more!

↓ Fold here second

Just fill out, fold, tape & drop this survey card into any mailbox!

BUSINESS REPLY MAIL
FIRST-CLASS MAIL　　PERMIT NO 71　　GROVER, MO

POSTAGE WILL BE PAID BY ADDRESSEE

**GREAT LAKES PRESS
PO BOX 550
WILDWOOD, MO 63040-9913**

NO POSTAGE
NECESSARY
IF MAILED
IN THE
UNITED STATES

Fold here first ↑

Cut along this line, then fold, tape and mail ✂